FE CIVIL EXAM PREP

All-in-One Study Guide with a Comprehensive Study Plan and 10 Full-Length Practice Tests! Includes 2500+ Practice Questions, Time-Saving Techniques and Problem-Solving Strategies

Jeremy Crainstone

© 2024 FE Civil Exam Prep All rights reserved. This book, 'FE Civil Exam Prep,' is intended for informational purposes only. Unauthorized copying, sharing, or any form of distribution of this book, in full or in part, is strictly forbidden. The publisher assumes no responsibility for any harm or loss arising from the use or misapplication of the information provided herein. This book is offered 'as is,' with no guarantees or assurances, whether express or implied. All trademarks and brand names cited are the property of their respective owners.

Table of Contents

Introduction 9
    Overview of the FE Civil Exam 11
    Navigating the NCEES FE Reference Handbook 12

Chapter 1: Mathematics and Statistics 14
    Analytic Geometry 14
        Coordinate Systems and Transformations 14
        Geometry Applications in Analytic Geometry 15
    Single-Variable Calculus 17
        Differentiation Essentials 17
        Integration Techniques and Applications 18
    Vector Operations 20
        Basics of Vectors 20
        Vector Geometry Applications 21
    Statistics 23
        Descriptive Statistics Essentials 23
        Inferential Statistics 25

Chapter 2: Ethics and Professional Practice 27
    Codes of Ethics in Professional Societies 27
    Professional Liability and Risk Mitigation 28
    Licensure and Professional Engineering Process 30
    Contracts and Contract Law Essentials 31

Chapter 3: Engineering Economics 33
    Time Value of Money 33
        Fundamental Concepts in Civil Engineering 33
        Advanced Financial Applications 34
    Cost Analysis 36
        Cost Types in Engineering Projects 36
        Cost Analysis 38
    Analyses 39
        Economic Evaluations: Break-Even and Cost-Benefit Analysis 39

    Sustainability & Life Cycle in Engineering 41

  Uncertainty 42

    Risk & Probability Fundamentals 42

    Decision Under Uncertainty 44

Chapter 4: Statics 46

  Resultants of Force Systems 46

    Force Components and Vector Addition 46

    Force System Applications 47

  Equivalent Force Systems 49

    Force-Couple Systems and Equivalence 49

    Applications of Equivalence in Engineering Structures 50

  Equilibrium of Rigid Bodies 51

    Equilibrium Conditions 51

    Applications of Equilibrium 53

  Frames and Trusses 55

    Analysis of Trusses: Joints and Sections Methods 55

    Frame Analysis: Force and Moment Distribution 57

  Centroid of Area 58

    Centroid Basics 58

    Composite Areas: Centroids of Shapes 60

  Area Moments of Inertia 61

    Basic Concepts: Moment of Inertia and Radius of Gyration 62

    Composite Shapes: Moments of Inertia Calculation 63

  Static Friction 65

    Friction Basics 65

    Friction in Inclined Planes and Wedges 66

Chapter 5: Dynamics 69

  Kinematics: Motion of Particles and Rigid Bodies 69

  Mass Moments of Inertia 70

  Force-Acceleration and Rigid Body Dynamics 71

  Work, Energy, and Power Principles 73

Chapter 6: Mechanics of Materials 75

Shear and Moment Diagrams 75

Diagram Basics: Shear and Moment Diagrams 75

Applications in Structural Design 76

Stresses and Strains 78

Stress Analysis Fundamentals 78

Thermal Stresses and Strain 79

Deformations 81

Axial and Torsional Deformations 81

Bending and Thermal Deformations 82

Combined Stresses and Mohr's Circle 84

Combined Stresses and Transformations 84

Mohr's Circle: Construction, Interpretation, Applications 85

Chapter 7: Materials 88

Mix Design of Concrete and Asphalt 88

Concrete Mix Design Essentials 88

Asphalt Mix Design Principles 89

Test Methods and Specifications 91

Concrete and Aggregates: Explain slump tests, compressive strength tests, gradation, and specific standards for aggregate properties. 91

Metals, Asphalt, and Wood: Discuss tensile tests for metals, asphalt binder tests (penetration/viscosity), and mechanical tests for wood. 92

Physical and Mechanical Properties 92

Basic Properties of Construction Materials 92

Strength Properties of Construction Materials 94

Chapter 8: Fluid Mechanics 96

Flow Measurement 96

Measurement Methods: Venturi, Orifice, Pitot Applications 96

Practical Measurement Techniques 97

Fluid Properties 99

Basic Fluid Properties 99

Specialized Fluid Properties in Engineering 100

Fluid Statics 101

Pressure in Static Fluids	101
Applications of Statics	103
Energy, Impulse, and Momentum of Fluids	104
Energy Principles in Fluid Systems	104
Impulse and Momentum in Fluid Dynamics	106

Chapter 9: Surveying 108

Angles, Distances, and Trigonometry	108
Measuring Techniques in Surveying	108
Applications of Triangulation and Error Adjustment	109
Area Computations	111
Area Calculations for Land Parcels	111
Advanced Applications in Area Estimation	112
Earthwork and Volume Computations	113
Cut and Fill: Earthwork Volume Computation	114
Advanced Volume Estimation Techniques	115
Coordinate Systems	117
Basics of Coordinate Systems	117
Transformations in Geospatial Engineering	118
Leveling	120
Differential Leveling Basics	120
Grade Computations in Topographic Surveying	122

Chapter 10: Water Resources & Environmental Engineering 124

Basic Hydrology	124
Hydrologic Processes in Water Resource Systems	124
Hydrology Applications	125
Basic Hydraulics	126
Flow Principles and Applications	126
Manning Equation and Channel Design	128
Pumps	130
Pump Types and Performance Basics	130
System Integration in Hydraulic Systems	132

- Water Distribution Systems 133
  - System Components in Water Distribution 133
  - Hydraulic Analysis: Pressure, Demand, Optimization 134
- Flood Control 136
  - Structural Methods for Flood Management 136
  - Flood Routing and Hydraulic Models 138
- Stormwater Management 139
  - Management Systems: Stormwater and Drainage Design 139
  - Water Quality and Pollutant Removal 141
- Collection Systems 143
  - Stormwater Collection Network Design 143
  - Wastewater Systems Design and Management 144
- Groundwater 147
  - Flow Principles and Groundwater Modeling 147
  - Wells and Drawdown Analysis 149
- Water Quality 150
  - Surface and Groundwater Quality Parameters 150
  - Basic Chemistry in Water Quality Assessment 152
- Testing and Standards 153
  - Water and Wastewater Testing 153
  - Air and Noise Standards 155
- Water and Wastewater Treatment 157
  - Drinking Water Treatment Methods 157
  - Wastewater Treatment Processes 159

Chapter 11: Structural Engineering 163
- Statically Determinant Structures Analysis 163
  - Structural Analysis Basics 163
  - Practical Applications: Analyzing Trusses, Beams, Frames 165
- Deflection of Statically Determinant Structures 166
  - Deflection Methods for Beams and Trusses 166
  - Frame Deflection Analysis 168

- Column Analysis 169
  - Buckling Basics: Euler's Formula and Critical Load 170
  - Boundary Conditions and Buckling Behavior 171
- Structural Determinacy and Stability Analysis 172
  - Determinate Structures: Criteria and Analysis 173
  - Stability Analysis 174
- Elementary Statically Indeterminate Structures 176
  - Indeterminate Beams and Frames: Superposition & Compatibility 176
  - Truss Redundancy Analysis 177
- Loads, Load Combinations, and Load Paths 179
  - Load Types in Structural Systems 179
  - Load Combinations and Design Codes 180
- Design of Steel Components 182
  - Steel Design Principles 182
  - Steel Connections: Bolted and Welded Joints 183
- Design of Reinforced Concrete Components 184
  - Concrete Design Basics 185
  - Column and Member Design 186

Chapter 12: Geotechnical Engineering 189
- Index Properties and Soil Classifications 189
  - Soil Properties: Indicators of Soil Behavior 189
  - Classification Systems: USCS and AASHTO Soils 190
- Phase Relations 192
  - Basic Soil Relationships 192
  - Weight-Volume Analysis in Engineering 193
- Laboratory and Field Tests 195
  - Laboratory Tests: Explain compaction, permeability, and shear strength tests performed in controlled settings. 195
  - Field Tests: Discuss SPT, CPT, and vane shear tests for determining in-situ soil properties. 196
- Effective Stress 196
  - Stress Basics in Soil Systems 196
  - Applications of Effective Stress 198

Stability of Retaining Structures 199

    Earth Pressure Theories: Rankine and Coulomb 200

    Wall Stability Analysis 201

Shear Strength 202

    Shear Strength Fundamentals 202

    Tests and Applications: Discuss direct shear, triaxial, and unconfined compression tests for shear strength evaluation. 204

Bearing Capacity 204

    Bearing Capacity of Shallow Foundations 204

    Factors Influencing Capacity 206

Foundation Types 207

    Shallow Foundations Design 207

    Deep Foundations: Piles and Drilled Shafts 209

Consolidation and Differential Settlement 210

    Consolidation and Settlement Analysis 210

    Settlement Effects and Mitigation Techniques 212

Slope Stability 213

    Slope Failure Mechanisms 213

    Stability Analysis for Embankments, Cuts, and Dams 215

Soil Stabilization 217

    Chemical Stabilization Additives 217

    Geosynthetics: Geotextiles and Geogrids 218

Chapter 13: Transportation Engineering 220

    Geometric Design 220

        Roadway Design Basics 220

        Intersection Design 221

    Pavement System Design 223

        Pavement Structure Essentials 223

        Rehabilitation Techniques 224

    Traffic Capacity and Flow Theory 226

        Flow Fundamentals in Traffic Analysis 226

        Capacity Analysis Using HCM Principles 227

- Traffic Control Devices 229
  - Traffic Signals and Signs 229
  - Road Markings and Intelligent Traffic Systems 230
- Transportation Planning 232
  - Travel Demand Modeling: Trip Generation & Distribution 232
  - Safety and Sustainable Transportation Strategies 233

Chapter 14: Construction Engineering 236
- Project Administration 236
  - Project Documentation: Contracts and Delivery Methods 236
  - Project Management Communication and Coordination 237
- Construction Operations and Methods 238
  - Safety Standards and Equipment Selection 238
  - Site Operations: Structures, Erosion, Productivity 240
- Project Controls 241
  - Scheduling and Tracking Techniques 241
  - Performance Metrics in Earned Value Analysis 243
- Construction Estimating 245
  - Cost Estimation Basics 245
  - Detailed Estimates: Contingencies and Cost Adjustments 246
- Interpretation of Engineering Drawings 248
  - Plan Reading: Understanding Construction Documents 248
  - Specifications and Symbols in Drawings 249

Conclusion 252
- Acknowledgment and Gratitude 252
- Encouragement for Future Growth 253

READ THIS PAGE CAREFULLY BEFORE YOU START PREPARING FOR THE EXAM

By purchasing this book, you gain access to a wealth of additional resources designed to help you prepare effectively. Don't think of your purchase as limited to the physical book in your hands—take advantage of all the extra resources available to you, such as:

- A study plan and topic checklist (crucial to ensure you cover everything required for the exam).
- In-depth theoretical explanations of key exam topics.
- Videos to help you master problem-solving techniques.
- Online and PDF practice test simulations.
- Tutorials on how to use the calculator efficiently and much more!

After consulting with many students who failed their exams, we identified a common issue: THESE STUDENTS WERE NOT ACCUSTOMED TO TAKING THE EXAM ONLINE OR USING THE REFERENCE HANDBOOK. This is a critical problem, as the exam is computer-based, meaning you must train under similar conditions to the actual exam. Otherwise, your chances of passing are significantly reduced.

FOR THIS REASON, WE CHOSE NOT TO INCLUDE PRACTICE EXERCISES IN THE BOOK but instead provide them in digital format. This approach will help you familiarize yourself with the digital exam format. While this may feel challenging at first, it greatly increases your chances of success.

Through conversations with numerous students, we learned the following:

- Most students who failed the exam reported focusing mainly on theory while neglecting practical exercises.
- Most students who passed the exam on their first attempt reported completing a large number of practical exercises and spending less time on theory.

WE RECOMMEND FOLLOWING THE BONUS RESOURCES IN THIS ORDER:

PHASE ONE

1. Scan the QR code and explore all the additional resources we have provided alongside this book.
2. Download the study plan to your computer.
3. Start watching the videos we've made available. Even if you feel lost at first, keep watching and focus on understanding the problem-solving methods demonstrated in each video.
4. Begin using the Reference Handbook immediately. It's essential to familiarize yourself with its structure and content.
5. Watch the calculator tutorials and memorize all the recommended strategies.

PHASE TWO

1. Choose which manual to begin with for each topic. (You'll find additional manuals among the supplementary resources, all of which are excellent for exam preparation.) You're not required to start with this book—you can use the others as well. This book is intentionally more concise and focused because your priority should be solving practical problems.

The key is to leverage all the supplementary materials to ensure comprehensive exam preparation. If you're short on study time or want to pass the exam quickly, we recommend focusing on the videos and additional resources, as they are specifically tailored to practical problems.

PHASE THREE

Start practicing problems yourself!

Happy studying!

SCAN THE QR CODE:

Introduction

Overview of the FE Civil Exam

The FE Civil Exam, administered by the National Council of Examiners for Engineering and Surveying (NCEES), is a pivotal step for engineering graduates aiming to secure a professional engineering license. This computer-based test (CBT) is meticulously designed to assess the examinee's knowledge and aptitude in civil engineering fundamentals, ensuring readiness for entry-level professional practice.

Structure: The exam encompasses 110 questions, covering topics that span the breadth of civil engineering. These questions are a mix of traditional multiple-choice and alternative item types (AITs), such as point and click, drag and drop, and fill-in-the-blank. This variety is intended to more accurately gauge the test-taker's ability to apply concepts rather than merely recall information.

Format: Questions are presented in a single, comprehensive session. The NCEES FE Reference Handbook is the only reference material allowed during the exam, available electronically. This ensures that all examinees have equal access to necessary resources, emphasizing understanding and application of principles over memorization.

Duration: The total appointment time is 6 hours, with approximately 5 hours and 20 minutes allocated to actual exam time. This includes a nondisclosed amount of unscored questions, used by NCEES for statistical purposes in maintaining exam quality and fairness. The remaining time is divided into a nondescript tutorial to familiarize candidates with the CBT interface before the exam begins, a scheduled 25-minute break after the first 55 questions, and brief surveys before and after the exam session.

Understanding the structure, format, and duration of the FE Civil Exam is crucial for effective preparation. Candidates are encouraged to familiarize themselves with the FE Reference Handbook and to develop a strategic study plan that covers all exam topics comprehensively.

This approach not only aids in passing the exam but also lays a solid foundation for a successful career in civil engineering.

Navigating the NCEES FE Reference Handbook

The **NCEES FE Reference Handbook** is the only resource allowed during the FE Civil Exam, making it an indispensable tool for candidates. This electronic handbook is designed to provide a comprehensive collection of engineering information and is accessible on the computer during the exam. The handbook covers all the subjects included in the exam, from mathematics and statistics to engineering economics, ethics, and professional practice, ensuring that candidates have quick access to formulas, tables, and conversion factors necessary for solving the exam questions.

To effectively utilize the **NCEES FE Reference Handbook** during the exam, candidates should familiarize themselves with its structure and content before the exam day. This involves understanding how the handbook is organized and where to find specific information quickly. The handbook is divided into sections corresponding to the major disciplines covered in the exam, with a detailed index and search function that allows for quick navigation.

Familiarity with the handbook can significantly reduce the time spent searching for information during the exam. Candidates are encouraged to practice using the electronic version of the handbook while studying and completing practice problems. This practice helps in developing a sense of where certain types of information are located within the handbook and how to use the search function effectively.

Effective navigation of the handbook involves mastering the use of bookmarks and the search feature. Keywords and common engineering terms can be used to find relevant sections quickly. Additionally, understanding the notation and conventions used in the handbook is crucial, as these may differ from those used in academic textbooks or class notes.

Candidates should also note that the **NCEES FE Reference Handbook** is periodically updated. It is essential to review the most current version available on the NCEES website to ensure familiarity with the latest standards, equations, and tables. The NCEES also provides a free PDF version of the handbook for candidates to download, allowing for offline study and reference.

Incorporating the handbook into study sessions not only aids in preparation but also builds confidence in using it as a reference tool under exam conditions. By becoming proficient in navigating the **NCEES FE Reference Handbook**, candidates can efficiently locate necessary information during the exam, saving time and reducing stress. This strategic approach to utilizing the handbook is a critical component of effective exam preparation and can contribute significantly to a candidate's ability to pass the FE Civil Exam on their first attempt.

Chapter 1: Mathematics and Statistics

Analytic Geometry

Coordinate Systems and Transformations

In the realm of analytic geometry, understanding the Cartesian coordinate system is foundational for analyzing the shapes and equations of various geometric figures such as circles, parabolas, ellipses, and hyperbolas. The Cartesian coordinate system facilitates the representation of geometric entities in a plane using ordered pairs (x, y). This system is instrumental in solving and graphing equations of geometric figures, enabling engineers to predict and model physical phenomena with precision.

Starting with the circle, its equation in the Cartesian plane is given by $(x-h)^2 + (y-k)^2 = r^2$, where (h, k) represents the center of the circle and r its radius. This equation is pivotal in civil engineering applications such as roundabout design, where understanding the curvature and dimensions of circular paths is essential.

Parabolas, with their characteristic U-shaped curves, are described by the equation $y = ax^2 + bx + c$ or $x = ay^2 + by + c$, depending on their orientation. The coefficients a, b, and c determine the parabola's width, direction, and vertex position, respectively. Parabolic trajectories are encountered in the analysis of projectile motion, optimizing the design of structures like bridges and arches to withstand loads efficiently.

Ellipses, elongated circles, are governed by the equation $\frac{(x-h)^2}{a^2} + \frac{(y-k)^2}{b^2} = 1$, where (h, k) is the center, and a and b are the semi-major and semi-minor axes, respectively. Ellipses model orbits in celestial mechanics, an analogy useful in understanding the trajectories of bodies under mutual gravitational attractions, relevant in geotechnical engineering for predicting soil movement around massive structures.

Hyperbolas, with their open curve structure, are represented by $\frac{(x-h)^2}{a^2} - \frac{(y-k)^2}{b^2} = 1$ or $\frac{(y-k)^2}{b^2} - \frac{(x-h)^2}{a^2} = 1$, depending on their orientation. The points (h, k) locate the center, while a and b define the distances to the vertices and foci, respectively. Hyperbolas are applicable in the analysis of wave propagation and signal processing, aiding in the design of acoustically optimized structures.

Transformations in the Cartesian plane, including translation, rotation, scaling, and reflection, are crucial for manipulating the position and orientation of geometric figures. Translation moves a figure by adding constant values to the x and y coordinates, effectively shifting the figure without altering its shape. Rotation involves pivoting a figure around a point, described by transformation matrices or specific rotation formulas, critical in structural analysis to assess the impact of forces from different directions. Scaling changes the size of a figure by multiplying the coordinates by a scale factor, a concept used in model scaling and simulating real-world scenarios in a controlled environment. Reflection flips a figure over a specified axis, mirroring its shape, an operation useful in symmetry analysis in structural design.

Equipping engineers with knowledge of these equations and transformations provides essential tools for modeling, analyzing, and designing structures and systems with precision and effectiveness. The capacity to convert physical problems into geometric representations and resolve them mathematically serves as a fundamental aspect of engineering problem-solving, enabling the creation of innovative solutions to intricate challenges encountered in civil engineering projects.

Geometry Applications in Analytic Geometry

In the realm of analytic geometry, the applications of geometric principles to solve real-world engineering problems are vast and varied. One of the foundational aspects of this discipline is the ability to calculate intersections, distances, midpoints, slopes, and areas of polygons, which are critical for the design, analysis, and implementation of civil engineering projects.

Intersections play a crucial role in determining the points at which two geometric figures, such as lines or curves, meet. The calculation of intersections involves solving systems of equations that represent each figure. For instance, the intersection of two lines can be found by solving their linear equations simultaneously. This concept is applied in traffic engineering to design and analyze road intersections, ensuring optimal flow and safety.

Distances between points in a Cartesian plane are calculated using the distance formula, $d = \sqrt{(x_2 - x_1)^2 + (y_2 - y_1)^2}$, where (x_1, y_1) and (x_2, y_2) are the coordinates of the two points. This formula is derived from the Pythagorean theorem and is essential for determining the length of segments in construction projects, such as the span between support beams or the length of piping required for utility networks.

Midpoints are found by averaging the coordinates of two points, given by the formula $\left(\frac{x_1 + x_2}{2}, \frac{y_1 + y_2}{2}\right)$. This calculation is particularly useful in dividing segments into equal parts for design symmetry or locating the center point for load distribution in structural engineering.

Slopes, defined as the ratio of the vertical change to the horizontal change between two points on a line, $\frac{y_2 - y_1}{x_2 - x_1}$, are fundamental in designing roads, ramps, and drainage systems. The slope is a measure of steepness and direction, critical for ensuring proper water runoff, accessibility, and vehicle safety on inclines.

Areas of polygons are calculated using various formulas depending on the polygon type. For rectangles and parallelograms, the area is the product of base and height. For triangles, it is half the product of base and height. For complex polygons, the area can be determined by dividing the shape into simpler figures or using the surveyor's formula, which involves the coordinates of the vertices. This knowledge is applied in land surveying to determine property boundaries, in architecture for floor planning, and in environmental engineering for assessing land use impact.

Each of these geometric applications is integral to the planning, design, and execution of civil engineering projects. By mastering these concepts, engineers can accurately model structures, predict outcomes, and devise solutions that are both functional and sustainable, addressing the analytical, determined, and goal-oriented nature of the engineering profession.

Single-Variable Calculus

Differentiation Essentials

In the realm of single-variable calculus, **differentiation** stands as a fundamental concept, pivotal for understanding the behavior of functions and their rates of change. This section delves into the core aspects of differentiation, including **limits**, **derivatives**, **rules of differentiation**, **tangent lines**, **optimization**, and **rates of change**, equipping readers with the necessary tools to tackle related problems on the FE Civil Exam with confidence.

Limits are the foundation upon which the concept of differentiation is built. They describe the behavior of a function as the input approaches a certain value. The limit of a function $f(x)$ as x approaches a is symbolized as $\lim_{x \to a} f(x)$. This concept is crucial for understanding derivatives, as it underpins the definition of the derivative itself.

The **derivative** of a function at a point provides the slope of the tangent line to the function at that point, representing the rate at which the function's value changes with respect to changes in the input value. Mathematically, the derivative of $f(x)$ with respect to x is denoted as $f'(x)$ or $\frac{df}{dx}$, and is defined as:

$$f'(x) = \lim_{h \to 0} \frac{f(x+h) - f(x)}{h}$$

This definition encapsulates the instantaneous rate of change of the function at any point x.

Rules of differentiation streamline the process of finding derivatives. Among these rules, the **Power Rule** ($d[x^n]/dx = nx^{n-1}$), **Product Rule** ($d[uv]/dx = u'v + uv'$), and **Quotient**

Rule ($d[u/v]/dx = (u'v - uv')/v^2$) are fundamental. Additionally, the **Chain Rule** is indispensable for finding the derivative of composite functions, expressed as $\dfrac{dy}{dx} = \dfrac{dy}{du} \cdot \dfrac{du}{dx}$.

Tangent lines are straight lines that touch a curve at a single point and have the same slope as the curve at that point. The equation of the tangent line to the curve $y = f(x)$ at a point a is given by:

$$y - f(a) = f'(a)(x - a)$$

This equation is pivotal for solving problems related to rates of change and optimization.

Optimization involves finding the maximum or minimum values of a function within a given domain. This is achieved by identifying critical points where the derivative is zero ($f'(x) = 0$) or undefined, and then determining whether these points correspond to maxima, minima, or saddle points by analyzing the second derivative or employing the First Derivative Test.

Rates of change describe how a quantity changes over time, a concept that is widely applied in engineering problems. The derivative provides a powerful tool for modeling and solving these problems, offering insights into the behavior of dynamic systems.

By mastering these concepts, readers will enhance their ability to solve a wide array of problems on the FE Civil Exam, leveraging differentiation to analyze and interpret the behavior of functions, optimize solutions, and accurately model rates of change in engineering contexts.

Integration Techniques and Applications

Integration, a cornerstone of calculus, serves as a fundamental tool in the analysis and solution of engineering problems, enabling the determination of areas under curves, volumes of solids, and the evaluation of various physical properties integral to civil engineering projects. This section delves into the concepts of definite and indefinite integrals, elucidating their applications and the basic techniques employed, such as substitution, to facilitate their computation.

The indefinite integral, represented as $\int f(x)\,dx$, signifies the collection of all antiderivatives of $f(x)$. It is essential to comprehend that the process of integration, in this context, is essentially the reverse operation of differentiation. If $F(x)$ is an antiderivative of $f(x)$, then the indefinite integral of $f(x)$ with respect to x is given by $F(x) + C$, where C represents the constant of integration. This constant embodies the infinite number of antiderivatives that can be generated by varying its value, highlighting the concept of a family of curves each shifted vertically from the others.

In contrast, the definite integral, denoted as $\int_a^b f(x)\,dx$, computes the net area between the curve $f(x)$ and the x-axis, across the interval $[a,b]$. This calculation is pivotal in various engineering applications, such as determining the displacement of an object given its velocity over time or calculating the total work done by a force. The Fundamental Theorem of Calculus bridges the concepts of differentiation and integration, stating that if $F(x)$ is an antiderivative of $f(x)$, then $\int_a^b f(x)\,dx = F(b) - F(a)$. This theorem not only simplifies the computation of definite integrals but also reinforces the intrinsic connection between the two principal operations of calculus.

The technique of substitution is frequently employed to simplify the process of integration, making it more tractable. This method involves replacing a part of the integrand with a new variable, thereby transforming the integral into a simpler form. For instance, consider the integral $\int x \cos(x^2)\,dx$. By letting $u = x^2$, we have $du = 2x\,dx$, or $\frac{1}{2}du = x\,dx$. The integral then becomes $\frac{1}{2}\int \cos(u)\,du$, which is more straightforward to evaluate. After integrating with respect to u, the substitution is reversed to express the result in terms of the original variable x.

The application of integration extends beyond the computation of areas and volumes to encompass a wide range of engineering problems, including those related to fluid dynamics, material strength, and thermal analysis. Mastery of integration techniques, therefore, is indispensable for aspiring civil engineers preparing for the FE Civil Exam. Through diligent study and practice, candidates can develop a robust understanding of integration, empowering them to tackle related exam questions with confidence and precision. This proficiency not only aids in exam preparation but also lays a solid foundation for future professional endeavors in the field of civil engineering, where the principles of calculus are applied to design, analyze, and solve real-world challenges.

Vector Operations

Basics of Vectors

Vectors are fundamental entities in engineering that encapsulate both magnitude and direction, making them indispensable in the analysis and solution of numerous problems in the field of civil engineering. A vector can be visually represented as an arrow, where the length denotes its magnitude, and the arrowhead points in the direction of the vector. The magnitude of a vector, denoted as $|\vec{v}|$ or $\|\vec{v}\|$, is a scalar quantity that represents the length of the vector and is always a non-negative value. It can be calculated using the Pythagorean theorem for a vector $\vec{v}$ with components v_x and v_y in two-dimensional space as $\|\vec{v}\| = \sqrt{v_x^2 + v_y^2}$.

The direction of a vector is typically described by the angle it makes with a reference axis, usually the positive x-axis in Cartesian coordinates. This angle can be determined using trigonometric functions, specifically the tangent function, where the angle θ is given by $\theta = \tan^{-1}\left(\frac{v_y}{v_x}\right)$.

Vector addition and subtraction are operations that follow the principles of parallelogram law and triangle law. For two vectors $\vec{a}$ and $\vec{b}$, their sum $\vec{c} = \vec{a} + \vec{b}$ is obtained by placing the tail of $\vec{b}$ at the head of $\vec{a}$ and drawing a vector from the tail of $\vec{a}$ to the head of $\vec{b}$. Subtraction follows a similar principle, where $\vec{c} = \vec{a} - \vec{b}$ is found by adding $\vec{a}$ to the negative of $\vec{b}$, which effectively reverses the direction of $\vec{b}$.

The dot product (or scalar product) and cross product (or vector product) are two types of multiplication involving vectors. The dot product of two vectors $\vec{a}$ and $\vec{b}$ with an angle θ between them is defined as $\vec{a} \cdot \vec{b} = |\vec{a}||\vec{b}|\cos(\theta)$, which results in a scalar. This operation is crucial in determining the angle between vectors and in projecting one vector onto another.

Conversely, the cross product of $\vec{a}$ and $\vec{b}$ is a vector $\vec{c}$ that is perpendicular to both $\vec{a}$ and $\vec{b}$, with a magnitude $|\vec{c}| = |\vec{a}||\vec{b}|\sin(\theta)$ and direction given by the right-hand rule. The cross product is vital in calculating torques and in determining the area of parallelograms formed by two vectors.

In the context of force and motion, vectors are used to represent forces acting on objects, velocities, and accelerations. The superposition principle allows for the combination of multiple forces acting at a point to be represented by a single resultant force vector. This simplification is instrumental in analyzing static equilibrium conditions and in solving problems related to the motion of objects, where the net force is the vector sum of all individual forces.

The fundamentals of vectors, along with their operations and applications, provide civil engineering professionals with essential tools for modeling and resolving intricate issues related to forces and motion. Proficiency in vector mathematics is vital for achieving success in the FE Civil Exam and serves as a critical basis for advanced study and professional practice in civil engineering, where vectors play a significant role in the analysis of structures, fluid dynamics, and various other systems.

Vector Geometry Applications

In the realm of vector geometry applications, understanding projections, angles between vectors, and the utilization of vector methods to solve geometry problems is paramount for engineers preparing for the FE Civil Exam. This section delves into these concepts, providing a comprehensive exploration to equip readers with the necessary tools for application in various engineering contexts.

Projections play a crucial role in vector geometry, especially in the analysis and solution of engineering problems. The projection of a vector $\vec{a}$ onto another vector $\vec{b}$ is a vector operation that results in a new vector $\vec{p}$ that lies on $\vec{b}$, representing the component of $\vec{a}$ in the direction of $\vec{b}$. Mathematically, the projection of $\vec{a}$ onto $\vec{b}$ is given by:

$$\vec{p} = \left(\frac{\vec{a} \cdot \vec{b}}{\|\vec{b}\|^2} \right) \vec{b}$$

This formula calculates the magnitude of $\vec{a}$ in the direction of $\vec{b}$ and scales $\vec{b}$ to this magnitude, yielding the projection vector $\vec{p}$. Projections are widely used in engineering to determine components of forces, velocities, and other vector quantities in specified directions, facilitating the analysis of systems in two and three dimensions.

Angles between vectors are another fundamental aspect of vector geometry, essential for understanding the orientation and relative direction of forces and other vector quantities in engineering problems. The angle θ between two vectors $\vec{a}$ and $\vec{b}$ can be found using the dot product formula:

$$\cos(\theta) = \frac{\vec{a} \cdot \vec{b}}{\|\vec{a}\|\|\vec{b}\|}$$

This equation derives the cosine of the angle from the dot product of $\vec{a}$ and $\vec{b}$ and the magnitudes of $\vec{a}$ and $\vec{b}$. By calculating the arccosine of the right-hand side, one can determine the angle θ, providing insights into the geometric and physical relationships between vectors in engineering applications.

Solving geometry problems using vector methods is a powerful approach that leverages the principles of vector addition, subtraction, dot and cross products, and projections. For instance, in determining the shortest distance between a point and a line or between two skew lines in three-dimensional space, vector methods offer a systematic and efficient solution. By representing lines as vector equations and points as position vectors, one can apply the cross product to find the area of parallelograms formed by vectors and use projections to find distances along specific directions. These techniques are invaluable in civil engineering for tasks such as analyzing the forces in truss systems, calculating moments in beams, and designing components to withstand specified loads.

Through the application of vector geometry in projections, understanding angles between vectors, and solving geometry problems with vector methods, engineers can enhance their analytical capabilities, enabling precise and efficient problem-solving in civil engineering projects. Mastery of these concepts is not only crucial for success on the FE Civil Exam but also forms a foundational skill set for professional practice, where vector geometry is applied to design, analyze, and optimize engineering solutions.

Statistics

Descriptive Statistics Essentials

Descriptive statistics serve as the cornerstone for analyzing data sets in engineering, providing a summary of the central tendency, dispersion, and shape of a dataset's distribution. Understanding these concepts is crucial for engineers to make informed decisions based on empirical data. This section delves into the fundamental aspects of descriptive statistics, including mean, median, mode, variance, standard deviation, histograms, and boxplots, equipping readers with the analytical tools needed to interpret and utilize data effectively in civil engineering contexts.

The **mean**, or average, represents the central tendency of a dataset and is calculated by summing all the values in the dataset and dividing by the number of values. Mathematically, for a dataset X with n values, the mean $\bar{x}$ is given by:

$$\bar{x} = \frac{1}{n} \sum_{i=1}^{n} x_i$$

where x_i represents the ith value in the dataset. The mean provides a quick snapshot of the dataset's overall level, but it can be skewed by outliers or non-symmetric distributions of data.

The **median** is the middle value of a dataset when it is ordered from least to greatest. If the dataset has an even number of observations, the median is the average of the two middle numbers. Unlike the mean, the median is not affected by outliers and provides a better measure of central tendency for skewed distributions.

The **mode** refers to the most frequently occurring value(s) in a dataset. A dataset may have one mode (unimodal), more than one mode (bimodal or multimodal), or no mode if all values are unique. The mode is particularly useful in identifying the most common or popular values in a dataset.

Variance measures the dispersion of a dataset by calculating the average of the squared differences from the Mean. The formula for the variance σ^2 of a dataset is:

$$\sigma^2 = \frac{1}{n}\sum_{i=1}^{n}(x_i - \bar{x})^2$$

Variance gives an idea of how much the data in a set are spread out from the mean. However, since it is in squared units of the data, it is often more practical to use the standard deviation.

Standard deviation is the square root of the variance and provides a measure of the spread of a dataset around the mean in the same units as the data. It is represented as σ and calculated as:

$$\sigma = \sqrt{\frac{1}{n}\sum_{i=1}^{n}(x_i - \bar{x})^2}$$

A low standard deviation indicates that the data points tend to be close to the mean, whereas a high standard deviation indicates that the data points are spread out over a wider range of values.

Histograms are graphical representations of the distribution of numerical data, where the data is divided into intervals or "bins," and the frequency of data points within each bin is depicted by the height of the bar. Histograms provide a visual interpretation of the underlying frequency distribution of a dataset, showing patterns that might not be evident from the raw data alone.

Boxplots, or box-and-whisker plots, offer a five-number summary of a dataset: the minimum, first quartile (Q1), median, third quartile (Q3), and maximum. These plots provide a visual summary of the central tendency, dispersion, and skewness of the dataset, as well as identifying outliers. The box represents the interquartile range (IQR), which contains the middle 50% of the data, and the whiskers extend to show the range of the data, excluding outliers.

By mastering these descriptive statistics tools, engineers can effectively summarize and analyze data, facilitating better decision-making and problem-solving in civil engineering projects. Understanding the properties of data through these statistical measures is essential for interpreting research findings, designing experiments, and implementing engineering solutions based on empirical evidence.

Inferential Statistics

Inferential statistics play a crucial role in engineering, particularly when making predictions or decisions based on data samples. This section delves into the core concepts of distributions, confidence intervals, hypothesis testing, regression, and curve fitting, providing a foundation for understanding how these tools can be applied in engineering analysis and FE Civil Exam preparation.

Distributions are fundamental in inferential statistics, describing how data points are spread out across different values. Engineers often encounter normal (Gaussian) distributions, which are characterized by the bell curve, but it's essential to recognize others like binomial and Poisson distributions. The choice of distribution depends on the nature of the data and the specific conditions of the engineering problem at hand. For example, the normal distribution is used for continuous data that tend to cluster around a mean value, while the binomial distribution applies to discrete data, representing the number of successes in a series of independent trials.

Confidence intervals provide a range of values within which the true population parameter is expected to fall, with a certain level of confidence. For instance, a 95% confidence interval for the mean compressive strength of a concrete batch indicates that, in 95 out of 100 similar samples, the mean strength will fall within this interval. Calculating a confidence interval involves determining the sample mean, the standard deviation, and the sample size, then using the appropriate z-score or t-score based on the sample size and desired confidence level.

Hypothesis testing is a systematic method used to decide whether to accept or reject a hypothesis about a population parameter, based on sample data. The process starts with stating the null hypothesis (H_0) and the alternative hypothesis (H_a), then calculating a test statistic from the sample data. Depending on the test (e.g., t-test, chi-square test), this statistic is compared to a critical value to determine whether to reject H_0. This method is pivotal in making informed decisions, such as determining if a new material has significantly different strength characteristics compared to an existing standard.

Regression analysis is a powerful statistical tool for modeling and analyzing the relationship between dependent and independent variables. Linear regression, for example, models the relationship in a linear manner, expressed as $y = mx + b$, where y is the dependent variable, x is the independent variable, m is the slope, and b is the y-intercept. Engineers use regression to predict outcomes, such as estimating the load capacity of a beam based on its dimensions and material properties.

Curve fitting involves finding the curve that best fits a series of data points, which is often used in conjunction with regression analysis. This process is crucial in engineering for modeling phenomena where the relationship between variables is not perfectly linear, requiring the use of polynomial or other non-linear models to accurately predict behavior under various conditions.

Chapter 2: Ethics and Professional Practice

Codes of Ethics in Professional Societies

Ethical principles form the cornerstone of professional conduct within the engineering field, guiding engineers in their responsibilities towards the public, their employers, and their peers. These principles are enshrined in the codes of ethics provided by professional and technical societies, such as the National Society of Professional Engineers (NSPE) and the American Society of Civil Engineers (ASCE). These codes serve as a framework for ethical decision-making and professional behavior, emphasizing the importance of integrity, honesty, fairness, and respect in all professional activities.

Responsibilities to the Public: Engineers have a paramount duty to safeguard the health, safety, and welfare of the public in their professional endeavors. This responsibility entails adhering to the highest standards of engineering practice and continuously updating their knowledge to incorporate the latest safety standards and regulations. When faced with situations that might endanger the public, engineers are ethically obligated to take appropriate action, which may include reporting the issue to authorities or the public if necessary.

Responsibilities to Employers and Clients: Loyalty, confidentiality, and honesty are key aspects of the engineer's responsibility to employers and clients. Engineers must act in the best interest of their employers and clients, provided that such actions do not conflict with their paramount duty to the public. Conflicts of interest should be avoided or disclosed to ensure transparency and maintain trust. Furthermore, engineers should not reveal confidential information without proper authorization or use such information for personal gain.

Conflicts of Interest: Engineers must be vigilant in identifying and managing conflicts of interest that may impair their ability to make impartial decisions. This includes avoiding situations where personal interests, relationships, or financial considerations have the potential to influence or appear to influence their professional judgment. When conflicts of interest cannot be avoided, they must be disclosed to affected parties.

Confidentiality: Protecting the confidentiality of information is a critical aspect of professional practice. Engineers are entrusted with sensitive information that, if disclosed improperly, could harm their clients, employers, or the public. Unauthorized disclosure of confidential information is considered unethical and can lead to legal consequences and damage to one's professional reputation.

Compliance with Laws and Regulations: Compliance with applicable laws, regulations, and standards is a fundamental ethical obligation. Engineers must not only comply with the letter of the law but also uphold the spirit of the laws governing their professional practice. This includes obtaining and maintaining the necessary licenses and certifications and adhering to the technical standards relevant to their work.

The codes of ethics of professional and technical societies provide a foundation for ethical engineering practice. By adhering to these codes, engineers demonstrate their commitment to professionalism, earning the trust and respect of the public, their employers, and their peers. Ethical behavior is not only a matter of personal integrity but also a critical component of professional excellence and the advancement of the engineering profession.

Professional Liability and Risk Mitigation

In the realm of engineering, **professional liability** encompasses the legal obligations that engineers have towards their clients and the public. This liability arises from the expectation that engineers will perform their duties with a certain level of care and expertise. Failure to meet these expectations can lead to **negligence**, which is defined as the failure to exercise the care that a reasonably prudent engineer would exercise under similar circumstances. Negligence can result in harm to individuals or damage to property, for which the engineer may be held legally responsible.

Duty of care is a fundamental concept in professional liability, establishing that engineers have a legal obligation to avoid actions or omissions that could foreseeably harm others. This duty extends to the design, inspection, and management of projects, requiring that all work is performed to the current standards and practices of the profession.

There are several **liability types** that engineers should be aware of:

1. **Direct Liability**: Occurs when the engineer's own actions or failures directly cause harm or damage.
2. **Vicarious Liability**: Arises when an engineer is held responsible for the actions of their employees or subcontractors.
3. **Strict Liability**: Applies in cases where activities inherently dangerous or under strict regulations cause harm, regardless of the engineer's negligence or intent.

To **mitigate risk** and manage professional liability, engineers can adopt several strategies:

- **Adherence to Professional Standards**: Keeping abreast of and adhering to the latest engineering standards, codes, and practices.
- **Continuous Education**: Engaging in lifelong learning to stay current with technological advancements and regulatory changes.
- **Documentation**: Maintaining thorough records of all decisions, calculations, and communications related to a project.
- **Communication**: Clearly communicating with clients, stakeholders, and team members about project risks, decisions, and changes.
- **Professional Liability Insurance**: Obtaining insurance to provide financial protection against claims of professional negligence.
- **Contractual Risk Management**: Using contracts to clearly define the scope of work, responsibilities, and limitations of liability.

The **consequences of professional misconduct or malpractice** can be severe, impacting an engineer's career, financial stability, and reputation. Legal actions can result in the awarding of damages to the injured party, disciplinary actions by professional licensing boards, and even criminal charges in cases of gross negligence. Beyond the legal ramifications, professional misconduct can erode public trust in the engineering profession, underscoring the importance of ethical practice and risk management.

Managing professional liability is crucial for engineers to protect themselves, their clients, and the public from the adverse consequences of professional negligence or misconduct. By adhering to ethical standards, maintaining a commitment to continuous improvement, and employing effective risk management strategies, engineers can uphold the integrity of their profession and contribute to the safety and well-being of society.

Licensure and Professional Engineering Process

The pathway to becoming a licensed professional engineer (PE) begins with passing the Fundamentals of Engineering (FE) exam, a critical step that assesses the candidate's comprehension of basic engineering principles. Administered by the National Council of Examiners for Engineering and Surveying (NCEES), the FE exam is designed for recent graduates and students who are close to finishing an undergraduate engineering degree. Following successful completion of the FE exam, candidates enter the Engineer-In-Training (EIT) or Engineering Intern (EI) phase, gaining valuable experience under the guidance of a seasoned PE.

Licensure Requirements vary by state but generally include a combination of education, experience, and examination components. Typically, a four-year degree from an ABET-accredited engineering program is required, followed by four years of progressive engineering experience under a PE. After meeting these prerequisites, candidates are eligible to sit for the Principles and Practice of Engineering (PE) exam, specific to their discipline.

Continuing Education plays a pivotal role in maintaining licensure. Most states mandate ongoing professional development to ensure PEs stay current with technological advancements, codes, and professional practices. This may include attending workshops, seminars, and conferences, or completing online courses and webinars.

Disciplinary Actions are enforced when professional conduct standards are violated. These can range from fines and license suspension to revocation, depending on the severity of the offense. Common violations include negligence, breach of contract, or failure to adhere to state laws and regulations governing professional engineering practice.

Benefits of Professional Engineering Licensure are manifold, enhancing not only the individual's career prospects but also public safety and welfare. Licensure distinguishes the engineer as a qualified professional, instilling confidence in employers, clients, and the public. It often leads to higher compensation, greater job opportunities, and the ability to bid for government contracts. Moreover, it signifies a commitment to ethical standards and lifelong learning, crucial attributes in the ever-evolving field of engineering.

To navigate the licensure process effectively, candidates must thoroughly understand the requirements set forth by their state's licensing board, including the specifics of the FE and PE exams. Preparation for these exams is paramount, as is a commitment to ethical practice and professional development throughout one's career. By achieving and maintaining licensure, engineers uphold the integrity of the profession, contributing to the advancement of society through the application of sound engineering principles.

Contracts and Contract Law Essentials

Contracts and contract law are fundamental aspects of professional engineering practice, ensuring that projects are executed within the bounds of agreed terms and legal requirements. Understanding the core elements of contracts and the nuances of contract law is essential for engineers to navigate the complexities of professional agreements and to safeguard their interests and those of their stakeholders.

Offer, Acceptance, and Consideration are the three primary elements that constitute a valid contract. An **offer** is a proposal by one party to enter into an agreement, presenting terms that are clear enough for acceptance. **Acceptance** must be an unequivocal agreement to the terms of the offer, communicated by the offeree to the offeror. **Consideration** refers to something of value that is exchanged between the parties, which can be a service, money, or an object. It is the reason or material benefit that compels the parties to enter into the contract.

Contracts can be classified into several **types** based on their nature and the specifics of the agreement:

1. **Fixed-Price Contracts**: These agreements involve a set price for the completion of the project or delivery of goods or services, regardless of the actual costs incurred.
2. **Cost-Reimbursement Contracts**: Under these contracts, the buyer reimburses the seller for allowable costs up to a predetermined limit, often including a fee or profit margin.

3. **Time and Materials Contracts (T&M)**: These agreements bill the client based on the actual time spent on the project and the materials used.

4. **Unit Price Contracts**: The payment is based on a per-unit rate for the work done or goods provided, useful in projects where the total quantities are uncertain.

A **breach of contract** occurs when one party fails to fulfill their obligations under the contract's terms, whether by not performing as promised, making it impossible for the other party to perform, or repudiating the contract (declaring they will not fulfill their obligations). The non-breaching party is entitled to seek legal remedies for breach of contract, which may include damages, specific performance, or cancellation and restitution.

The **roles and responsibilities** in project agreements are clearly defined within the contract, outlining what each party is expected to do. This includes the scope of work, timelines, payment schedules, and standards for the quality of work. Engineers, as part of their professional practice, must ensure that contracts they enter into or manage are clear on the expectations and obligations of all parties involved. This clarity helps prevent disputes and provides a basis for resolving any issues that may arise.

In managing contracts, engineers must also be aware of the **legal implications** of their actions and decisions. This includes understanding how changes to the project scope, delays, and other unforeseen events are addressed within the contract framework. Effective contract management involves regular communication between parties, meticulous record-keeping, and a proactive approach to identifying and resolving potential issues before they escalate into disputes.

In the realm of engineering, where projects often involve significant investments of time, money, and resources, a solid understanding of contracts and contract law is indispensable. It not only protects the legal and financial interests of the engineer and their stakeholders but also contributes to the smooth execution and successful completion of engineering projects.

Chapter 3: Engineering Economics

Time Value of Money

Fundamental Concepts in Civil Engineering

Understanding the time value of money is foundational to engineering economics, as it allows engineers to assess the worth of future cash flows in today's dollars. This concept is predicated on the principle of equivalence, which posits that different sums of money at different times can be equivalent in value if they consider the interest rate, or the cost of capital. The interest rate is a critical factor in the time value of money, representing the rate of return required to make an investment worthwhile. It can be expressed as an annual percentage rate (APR) and influences how future cash flows are discounted back to their present value.

Present worth (or present value) is a key concept that quantifies the current value of a future cash flow or series of cash flows, discounted at a specific interest rate. The formula for calculating the present worth of a single future amount is given by:

$$PV = \frac{FV}{(1+i)^n}$$

where PV is the present value, FV is the future value, i is the interest rate per period, and n is the number of periods. This formula helps in comparing the value of money received at different times by converting future values into their equivalent present values.

Future worth (or future value) is the opposite of present worth, representing the amount that a present sum of money will grow to at a future date, given a specific interest rate. The future worth can be calculated using the formula:

$$FV = PV \times (1+i)^n$$

where FV is the future value, PV is the present value, i is the interest rate per period, and n is the number of periods. This calculation is crucial for understanding how investments grow over time.

Cash flow diagrams are visual representations used to illustrate the inflows and outflows of cash over a period. These diagrams are essential tools in engineering economics for visualizing the timing and magnitude of cash flows associated with projects or investments. A typical cash flow diagram consists of a horizontal line representing time, with arrows pointing upwards for cash inflows and downwards for cash outflows. The position of each arrow along the time axis indicates the timing of the cash flow, while the length or label of the arrow denotes the magnitude. Cash flow diagrams serve as the basis for further economic analysis, including the calculation of present and future worth.

The concepts of equivalence, interest rates, present worth, future worth, and cash flow diagrams play a crucial role in engineering economics, equipping engineers with essential tools for financial decision-making in engineering projects. These principles allow engineers to evaluate the economic feasibility of projects, assess various investment alternatives, and make well-informed choices that reflect the time value of money.

Advanced Financial Applications

In the realm of engineering economics, understanding advanced financial applications such as the rate of return, equivalent annual worth, capitalized cost, and amortization is crucial for making informed decisions on investments and project evaluations. These concepts extend the foundational principles of the time value of money, providing engineers with the tools to assess the financial viability and performance of engineering projects over their lifecycle.

The rate of return, often denoted as i, is a critical measure that represents the profitability of an investment. It is calculated as the ratio of the annual net profit (or loss) to the initial investment cost, expressed as a percentage. The formula for the rate of return is given by:

$$i = \frac{\text{Annual Net Profit}}{\text{Initial Investment Cost}} \times 100\%$$

This metric is pivotal in comparing the efficiency of different investments or projects, where a higher rate of return signifies a more desirable outcome.

Equivalent annual worth (EAW) is another vital concept that converts the net present value of all cash flows associated with a project into an equivalent uniform annual value over the project's lifespan. This is particularly useful when comparing projects with different durations and cash flow patterns. The formula to calculate EAW is:

$$EAW = \frac{NPV \times i}{1 - (1+i)^{-n}}$$

where NPV is the net present value of the project's cash flows, i is the interest rate, and n is the number of periods. EAW allows for a uniform comparison basis, facilitating the selection of the most economically viable project.

Capitalized cost represents the present worth of a project or asset that has infinite life. This concept is essential when evaluating investments in infrastructure or projects with long-term benefits. The capitalized cost is calculated by dividing the annual operating cost by the interest rate, as shown in the formula:

$$\text{Capitalized Cost} = \frac{\text{Annual Operating Cost}}{i}$$

This calculation provides a snapshot of the total cost of owning and operating an asset, aiding in the financial planning and budgeting process.

Amortization, in the context of engineering economics, refers to the process of gradually reducing the value of an intangible asset or paying off a debt over a period. For assets, amortization spreads the cost of the asset over its useful life, reflecting its consumption or the pattern of its economic benefits. For debts, it involves making regular payments that cover both principal and interest, eventually bringing the balance to zero. The formula for the amortization payment (A) of a loan is:

$$A = P \frac{i(1+i)^n}{(1+i)^n - 1}$$

where P is the principal amount borrowed, i is the interest rate per period, and n is the total number of payments. This systematic reduction or allocation helps in managing finances effectively, ensuring that assets and liabilities are accurately represented over time.

These advanced financial applications empower engineers to perform comprehensive economic analyses, ensuring that projects not only meet technical and operational requirements but also adhere to financial sustainability and efficiency criteria. By applying these concepts, engineers can optimize resource allocation, enhance project outcomes, and contribute to the economic success of their organizations.

Cost Analysis

Cost Types in Engineering Projects

In the realm of engineering economics, understanding the various types of costs associated with projects is crucial for effective financial planning and analysis. These costs can be categorized into fixed, variable, direct, indirect, incremental, and sunk costs, each playing a significant role in the economic evaluation of engineering projects.

Fixed costs are expenses that do not change with the level of production or service delivery within a certain range of activity and over a specified period. These costs are independent of the output level, meaning they remain constant whether the project is in a dormant state or operating at full capacity. Examples of fixed costs in engineering projects include salaries of permanent staff, lease payments for equipment or facilities, and insurance premiums. For instance, the monthly rent for a construction site office is a fixed cost, as it does not vary with the amount of construction work being carried out.

Variable costs, in contrast, fluctuate directly with the level of production or service delivery. These costs increase as output levels rise and decrease as output levels fall. Materials, labor, and energy consumption are typical examples of variable costs in engineering projects. For example, the cost of concrete for a bridge construction project would be a variable cost, as the amount of concrete required would directly depend on the size and specifications of the bridge.

Direct costs are those that can be directly attributed to a specific project, product, or service. These costs are easily identifiable and can be allocated to a particular project without ambiguity. Direct costs include expenses such as labor specifically hired for the project, materials, and subcontracting costs. For an engineering project like the construction of a highway, direct costs would encompass the cost of asphalt, wages of construction workers, and fees paid to subcontractors for electrical installations.

Indirect costs, also known as overheads, cannot be directly linked to a specific project but are necessary for the operation of the business as a whole. These costs include utilities, administrative salaries, and security services. Indirect costs are allocated to projects based on various allocation bases such as labor hours, machine hours, or material costs. For example, the salary of the project manager overseeing multiple projects would be considered an indirect cost, as their work benefits more than one project.

Incremental costs, or marginal costs, refer to the additional costs incurred when increasing the output level or scope of a project. These costs are relevant for decision-making when evaluating the financial implications of expanding a project's scope or enhancing its output. An example of an incremental cost is the additional expense of constructing an extra mile of highway beyond the initially planned length.

Sunk costs represent expenses that have already been incurred and cannot be recovered. These costs should not influence future project decisions since they remain constant regardless of the outcome. An example of a sunk cost in engineering projects is the expenditure on feasibility studies conducted before the project's approval. Once spent, these costs do not affect the future financial viability of the project and should not be considered in future decision-making processes.

Enabling engineers and project managers to accurately analyze financial aspects, prepare budgets, control costs, and make informed decisions enhances the economic viability and success of engineering projects. Categorizing costs effectively allows engineers to identify areas for cost optimization, which contributes to the overall efficiency and sustainability of engineering endeavors.

Cost Analysis

Average cost and marginal cost are pivotal concepts in the realm of engineering economics, especially when it comes to cost analysis and resource allocation in engineering projects. The average cost, often referred to as the unit cost, is calculated by dividing the total cost (TC) by the quantity (Q) of output produced. This can be represented by the formula:

$$AC = \frac{TC}{Q}$$

This metric is crucial for understanding the overall cost efficiency of a project or operation. It provides insight into the cost of producing one unit of output, allowing engineers and project managers to gauge the economic scale of operations. For instance, in a construction project, understanding the average cost of laying one square foot of asphalt helps in budgeting and pricing services.

Marginal cost, on the other hand, represents the cost of producing one additional unit of output. It is derived from the change in total cost that arises from producing one additional unit of a product or service. Mathematically, it is expressed as the derivative of the total cost (TC) with respect to the quantity (Q), or the change in TC divided by the change in Q. The formula for marginal cost (MC) is given by:

$$MC = \frac{\Delta TC}{\Delta Q}$$

Marginal cost is a critical factor in decision-making processes, particularly when determining the optimal level of production. It helps in identifying the point at which the cost of producing one more unit equals the revenue it generates, known as the break-even point. This concept is instrumental in optimizing production levels to ensure profitability.

Cost estimation methods are essential tools for engineers to predict the costs associated with a project accurately. These methods range from simple analogies based on past projects to complex parametric equations that consider various project variables. One common approach is the use of historical data, where costs are estimated based on the actual costs of similar projects, adjusted for any differences in scale, complexity, or location. Another method is the bottom-up approach, where the total project cost is estimated by summing the costs of all individual activities or components. This method is more detailed and often more accurate but requires a thorough understanding of the project scope and design.

Resource allocation in engineering projects involves distributing available resources, such as labor, materials, and capital, in a manner that maximizes efficiency and effectiveness. Effective resource allocation is crucial for meeting project deadlines, staying within budget, and achieving the desired quality standards. It requires a comprehensive understanding of the project scope, tasks, and schedules, as well as the ability to anticipate potential bottlenecks and constraints. Techniques such as Critical Path Method (CPM) and Program Evaluation and Review Technique (PERT) are commonly used to optimize the allocation of resources by identifying the critical tasks that directly impact the project timeline.

In the context of engineering economics, understanding and applying the concepts of average cost, marginal cost, cost estimation methods, and resource allocation are fundamental to the successful financial management of engineering projects. These concepts enable engineers to make informed decisions that balance cost, time, and quality, ultimately contributing to the economic viability and success of their projects.

Analyses

Economic Evaluations: Break-Even and Cost-Benefit Analysis

Economic evaluations are critical in engineering economics, providing a quantitative basis for comparing the financial viability of different projects or decisions. Two fundamental tools in this process are **break-even analysis** and **benefit-cost ratio** (BCR), both of which aid in making informed decisions based on economic feasibility.

Break-even analysis is a method used to determine the point at which a project or business will become profitable. It calculates the volume of production or sales at which total revenues equal total costs, known as the break-even point (BEP). The formula for finding the BEP in terms of units is given by:

$$\text{BEP (units)} = \frac{\text{Fixed Costs}}{\text{Price per Unit} - \text{Variable Cost per Unit}}$$

where **Fixed Costs** are costs that do not change with the level of output, **Price per Unit** is the selling price of the product, and **Variable Cost per Unit** is the cost that varies with the level of production. This analysis is crucial for understanding the minimum output necessary to cover costs, thereby informing decisions on whether a project or venture is financially viable.

The **benefit-cost ratio (BCR)**, on the other hand, is a financial metric that compares the benefits of a decision or project to its costs, expressing the relationship as a ratio. A BCR greater than 1.0 indicates that the project's benefits outweigh its costs, suggesting economic feasibility. The formula for BCR is:

$$\text{BCR} = \frac{\text{Present Value of Benefits}}{\text{Present Value of Costs}}$$

Calculating the present value of benefits and costs involves discounting future cash flows back to their present value, using a discount rate that reflects the project's cost of capital or minimum rate of return. This rate is crucial as it accounts for the time value of money, ensuring that future cash flows are appropriately valued in today's terms.

Both break-even analysis and BCR are instrumental in guiding decision-making processes. They provide a clear, quantitative foundation for comparing projects, assessing their economic viability, and ultimately selecting the most financially sustainable options. In the context of engineering economics, these tools enable engineers to evaluate the economic impact of their projects and decisions, ensuring that resources are allocated efficiently and effectively to maximize financial returns.

Furthermore, these analyses support strategic planning by highlighting the financial implications of various scenarios, allowing engineers to anticipate and mitigate risks associated with cost overruns, revenue shortfalls, and changes in market conditions. By integrating break-even analysis and BCR into the economic evaluation process, engineers can enhance the financial sustainability of their projects, contributing to the overall success and profitability of their organizations.

Sustainability & Life Cycle in Engineering

In the realm of engineering economics, understanding the principles of sustainability and life cycle cost is paramount for developing engineering solutions that are not only economically viable but also environmentally responsible. The concept of life cycle cost (LCC) encompasses all costs associated with the life span of an asset, including initial design and construction costs, operation, maintenance, and the eventual disposal costs. This comprehensive approach ensures that engineers and decision-makers consider the total expenditure over an asset's lifetime, rather than just the upfront costs, leading to more sustainable economic decisions.

The LCC analysis is instrumental in evaluating the cost-effectiveness of incorporating renewable energy technologies into engineering projects. By assessing the initial investment against the long-term savings in energy costs and environmental benefits, engineers can justify the higher upfront costs associated with renewable energy systems. For instance, the installation of solar panels or wind turbines may have a significant initial cost, but when analyzed over their operational life, the reduction in energy costs and the environmental benefits they provide often outweigh the initial investment. This is particularly relevant in the context of rising environmental concerns and the global push towards reducing carbon footprints.

Moreover, sustainability in engineering solutions extends beyond economic and environmental considerations to include social aspects as well. Sustainable engineering practices aim to meet the current generation's needs without compromising the ability of future generations to meet their own. This involves the careful selection of materials, energy sources, and technologies that minimize environmental impact and promote social well-being. For example, choosing materials with lower embodied energy and higher recyclability contributes to more sustainable construction practices.

The economics of renewable energy also play a crucial role in sustainable engineering. As the cost of renewable energy technologies continues to decrease due to advancements in technology and increased production scale, the economic feasibility of these options becomes more attractive. The economic evaluation of renewable energy projects often involves calculating the levelized cost of energy (LCOE), which represents the per-unit cost of building and operating a generating plant over an assumed financial life and output. A lower LCOE indicates a more cost-effective energy generation option, making renewable energy increasingly competitive with traditional fossil fuel sources.

Incorporating sustainability and life cycle cost considerations into engineering solutions requires a multidisciplinary approach that balances economic, environmental, and social factors. Engineers must employ advanced analytical methods to assess the viability and impact of their designs comprehensively. This includes the use of software tools for life cycle assessment (LCA) and cost-benefit analysis (CBA) to quantitatively evaluate the sustainability of engineering projects. By integrating these considerations into the decision-making process, engineers can contribute to the development of solutions that are not only economically viable but also environmentally sustainable and socially responsible.

The shift towards sustainable engineering practices is driven by a combination of regulatory requirements, societal expectations, and the intrinsic motivation to preserve the planet for future generations. As such, the principles of sustainability and life cycle cost are becoming increasingly integral to the engineering profession, guiding the development of innovative solutions that address the complex challenges of the 21st century. Through the diligent application of these principles, engineers play a critical role in shaping a sustainable and innovative future, demonstrating that economic development and environmental stewardship can go hand in hand.

Uncertainty

Risk & Probability Fundamentals

In the realm of engineering economics, the concepts of risk and probabilities are pivotal for making informed decisions under uncertainty. Risk quantification and the understanding of probability distributions are tools that enable engineers to anticipate and mitigate potential setbacks in projects, ensuring that decisions are both economically viable and resilient to unforeseen challenges.

The expected value, denoted as $E(X)$, is a fundamental concept in probability theory and serves as a cornerstone in the analysis of risk. It represents the long-term average outcome of a random variable over a large number of trials and is calculated as the weighted average of all possible values. For a discrete random variable X with possible values x_i and corresponding probabilities $P(x_i)$, the expected value is given by:

$$E(X) = \sum_{i=1}^{n} x_i P(x_i)$$

This formula encapsulates the essence of expected value by multiplying each outcome by its probability and summing these products. In the context of engineering economics, the expected value offers a method to quantify the average expected outcome of different financial decisions, such as the return on investment in various projects, taking into account the probability of each outcome.

Probability distributions, on the other hand, provide a systematic way to describe the likelihood of various outcomes. They are categorized into discrete and continuous distributions, with common examples including the binomial distribution for discrete variables and the normal distribution for continuous variables. The choice of distribution depends on the nature of the random variable being analyzed.

The binomial distribution, for instance, is used when the outcome of an event can be classified into two categories (success or failure), and the event is repeated a fixed number of times. The probability mass function (PMF) of a binomial distribution is given by:

$$P(X = k) = \binom{n}{k} p^k (1-p)^{n-k}$$

where n is the number of trials, k is the number of successful outcomes, p is the probability of success on a single trial, and $\binom{n}{k}$ is the binomial coefficient.

The normal distribution, characterized by its bell-shaped curve, is widely used to model continuous random variables that are influenced by a large number of small, independent effects. The probability density function (PDF) of a normal distribution is:

$$f(x) = \frac{1}{\sigma\sqrt{2\pi}} e^{-\frac{1}{2}\left(\frac{x-\mu}{\sigma}\right)^2}$$

where μ is the mean, σ is the standard deviation, and e is the base of the natural logarithm. The normal distribution is particularly useful in engineering economics for modeling uncertainties in project costs, demand forecasts, and other variables that affect economic analysis.

Basic risk quantification involves the application of these probabilistic concepts to evaluate the potential variability in outcomes and their implications for decision-making. By calculating the expected value and analyzing the probability distribution of project returns, engineers can identify the level of risk associated with different options and choose strategies that optimize the balance between risk and reward. This process often involves scenario analysis, sensitivity analysis, and the use of decision trees to explore and evaluate the outcomes of different decision paths under uncertainty.

Decision Under Uncertainty

In the realm of engineering economics, decision-making under uncertainty is a critical skill that engineers must master to ensure the viability and success of their projects. Sensitivity analysis, risk mitigation strategies, and scenario planning are essential tools in this process, enabling engineers to navigate the complexities of uncertain environments effectively.

Sensitivity Analysis is a technique used to determine how different values of an independent variable affect a particular dependent variable under a given set of assumptions. This analysis is crucial in engineering economics for identifying the most influential factors that could impact the outcome of a project. By systematically varying key parameters one at a time while keeping other variables constant, engineers can pinpoint which variables have the most significant effect on the project's economic feasibility. The mathematical representation of sensitivity analysis can be expressed as the partial derivative of the outcome variable Y with respect to an input variable X_i, denoted as $\frac{\partial Y}{\partial X_i}$. This calculation helps in understanding how changes in X_i will affect Y, providing a basis for more informed decision-making.

Risk Mitigation Strategies involve identifying potential risks, assessing their impact, and implementing measures to manage or mitigate these risks. In engineering economics, risks can range from cost overruns and delays to technological failures and market fluctuations. Effective risk mitigation requires a comprehensive understanding of the project's scope and environment, followed by the development of a risk matrix that categorizes risks based on their probability and impact. Strategies such as diversification, risk transfer through insurance, contingency planning, and adopting flexible design principles can help manage risks. For instance, establishing a contingency fund calculated as a percentage of the project's total cost, typically between 5% and 20%, depending on the risk level, can provide a financial buffer against unforeseen expenses.

Scenario Planning is a strategic planning method that allows engineers to explore and prepare for multiple future conditions. It involves creating detailed narratives about different ways the future could unfold and analyzing how these scenarios could impact project outcomes. Scenario planning is particularly useful in long-term projects where future conditions are highly uncertain. By considering a range of possible futures, engineers can develop flexible strategies that are robust across various scenarios. The process includes identifying critical uncertainties, developing plausible scenarios, evaluating these scenarios' implications, and integrating findings into project planning and decision-making. For example, in the context of a construction project, scenarios might include changes in regulatory policies, economic downturns, or advancements in technology, each requiring different responses to ensure project success.

Employing these tools allows engineers to make more resilient and informed decisions under uncertainty, enhancing the likelihood of project success. By understanding the sensitivity of projects to various factors, implementing strategies to mitigate identified risks, and planning for multiple future scenarios, engineers can navigate the complexities of their projects with greater confidence and precision. These methodologies not only contribute to the economic viability of projects but also to their sustainability and adaptability in the face of changing conditions.

Chapter 4: Statics

Resultants of Force Systems

Force Components and Vector Addition

In the realm of statics, understanding how to resolve forces into their components is fundamental for analyzing the equilibrium of structures and mechanical systems. This process involves breaking down a force into its horizontal and vertical components, which simplifies the analysis of systems subjected to multiple forces. The resolution of forces is particularly crucial when dealing with 2D and 3D force systems, as it allows engineers to apply the principles of equilibrium more effectively.

For a force $\mathbf{F}$ acting at an angle θ to the horizontal, the horizontal (F_x) and vertical (F_y) components can be determined using trigonometric relationships:

$$F_x = F \cos(\theta)$$

$$F_y = F \sin(\theta)$$

These components are vectors and can be represented graphically or calculated numerically. In 2D systems, this resolution of forces into F_x and F_y simplifies the application of the equations of equilibrium, $\sum F_x = 0$ and $\sum F_y = 0$, which state that the sum of the horizontal forces and the sum of the vertical forces in a system in equilibrium must each equal zero.

In 3D systems, the resolution of forces becomes slightly more complex due to the addition of a third dimension. A force $\mathbf{F}$ in 3D space can be resolved into its components along the x, y, and z axes. If α, β, and γ are the angles that $\mathbf{F}$ makes with the x-, y-, and z-axes, respectively, the components of $\mathbf{F}$ can be expressed as:

$$F_x = F \cos(\alpha)$$

$$F_y = F \cos(\beta)$$

$F_z = F\cos(\gamma)$

It's important to note that $\cos^2(\alpha) + \cos^2(\beta) + \cos^2(\gamma) = 1$, a relationship that arises from the directional cosines of a vector in 3D space.

Vector addition is another critical concept in the analysis of force systems. When multiple forces act on a body, the total effect is equivalent to the effect of a single force known as the resultant force. This resultant can be found by vectorially adding the individual forces. For two forces $\mathbf{F}_1$ and $\mathbf{F}_2$ acting on a point, the resultant $\mathbf{R}$ can be determined using the parallelogram law of vector addition:

$$\mathbf{R} = \mathbf{F}_1 + \mathbf{F}_2$$

Graphically, $\mathbf{R}$ is the diagonal of the parallelogram formed by $\mathbf{F}_1$ and $\mathbf{F}_2$. Analytically, if $\mathbf{F}_1$ and $\mathbf{F}_2$ are expressed in terms of their i, j, and k components (for 3D systems), $\mathbf{R}$ can be calculated by summing these components:

$$\mathbf{R} = (F_{1x} + F_{2x})\mathbf{i} + (F_{1y} + F_{2y})\mathbf{j} + (F_{1z} + F_{2z})\mathbf{k}$$

This approach to vector addition and the resolution of forces into components are foundational techniques in statics, enabling engineers to analyze and design stable structures and mechanical systems by ensuring that all forces and moments are in equilibrium.

Force System Applications

In the analysis of static systems, understanding the applications of force systems, including concurrent, parallel, and distributed forces, alongside the calculation of moment resultants, is crucial for engineering professionals. This knowledge is foundational for designing and evaluating the stability of structures and mechanical systems under various loading conditions.

Concurrent forces refer to a set of forces that act through a common point, even if their lines of action do not necessarily lie along the same line. The resultant of concurrent forces can be determined by vector addition, where each force is represented by a vector in terms of its magnitude and direction. The algebraic sum of the horizontal and vertical components of these forces gives the components of the resultant force. Mathematically, if forces $\mathbf{F}_1, \mathbf{F}_2, \ldots, \mathbf{F}_n$ act concurrently at a point, and θ_i is the angle that $\mathbf{F}_i$ makes with a reference axis, the resultant force $\mathbf{R}$ in vector form is given by:

$$\mathbf{R} = \sum_{i=1}^{n} \mathbf{F}_i = \left(\sum_{i=1}^{n} F_i \cos(\theta_i) \right) \mathbf{i} + \left(\sum_{i=1}^{n} F_i \sin(\theta_i) \right) \mathbf{j}$$

Parallel forces are forces that have the same line of action but may act in the same or opposite directions. The resultant of parallel forces can be found by summing the magnitudes of the forces acting in the same direction and subtracting the magnitudes of forces acting in the opposite direction. The location of the resultant force, or the line of action, can be determined using the principle of moments, where the moment of the resultant force about any point is equal to the sum of the moments of the individual forces about the same point.

Distributed forces are forces that act over a certain area or along a line, such as the weight of a beam acting along its length or the pressure exerted by a fluid on a surface. These forces are often represented as uniformly distributed loads (UDL), varying distributed loads (VDL), or concentrated loads for simplicity in analysis. The equivalent resultant force of a distributed load is equal to the area under the load distribution curve, and its point of application, known as the centroid, is located at the center of gravity of this area. For a uniformly distributed load w acting over a length L, the resultant force F is wL, and it acts at the midpoint of L.

Moment resultants are moments caused by forces about a specific point or axis and are crucial for understanding the rotational effects of forces on bodies. The moment M of a force F about a point is given by the product of the force magnitude and the perpendicular distance d from the point to the line of action of the force, expressed as $M = Fd$. For systems with multiple forces, the total moment about a point is the algebraic sum of the moments of the individual forces about that point. This principle is essential for analyzing the equilibrium of beams, frames, and other structural elements where rotational stability is a concern.

In the context of statics, the ability to decompose complex force systems into simpler, more analyzable components such as concurrent, parallel, and distributed forces, and to calculate the resultant forces and moments, is indispensable. This approach not only simplifies the analysis but also enhances the accuracy and efficiency of designing safe and reliable engineering structures and systems.

Equivalent Force Systems

Force-Couple Systems and Equivalence

In the realm of statics, understanding **force-couple systems** is crucial for simplifying complex force configurations into more manageable forms. This section delves into the principles of force-couple equivalence, the methodology for combining forces and moments, and the strategies for system reduction, which are pivotal for analyzing engineering structures efficiently.

Force-couple equivalence is a fundamental concept that allows engineers to replace a system of forces acting on a body with a simpler equivalent system without altering the external effects on the body. This equivalence is established when the resultant force and the resultant moment of the original system about any point are equal to those of the simplified system. Mathematically, if a force $\mathbf{F}$ acts at a point A on a body, it can be replaced by the same force $\mathbf{F}$ acting at any other point B along its line of action, combined with a couple $M = \mathbf{F} \times d$, where d is the perpendicular distance between the lines of action of the original and relocated forces.

To **combine forces and moments**, it's essential to understand that forces can be added vectorially, and moments, being free vectors, can be summed algebraically about any point. When dealing with multiple forces and couples acting on a body, the first step is to find the resultant force $\mathbf{F}_R$ by vector addition of all forces. Concurrently, calculate the resultant moment M_R about a chosen point by summing the moments of all forces about that point, including the moments of any existing couples. The combined effect of $\mathbf{F}_R$ and M_R represents the equivalent force-couple system.

System reduction involves simplifying a complex force system into an equivalent force-couple system at a chosen point, which significantly eases the analysis. This process entails:
1. Determining the resultant force $\mathbf{F}_R$ by summing all forces acting on the body.

2. Calculating the resultant moment M_R about a chosen point, considering both the moments produced by the forces and any initial couples present.

3. Representing the simplified system by the resultant force F_R applied at the chosen point, accompanied by the resultant moment M_R.

This approach not only streamlines the analysis of static equilibrium but also facilitates the design and evaluation of structures by reducing complex systems to more tractable models. Understanding and applying the principles of force-couple equivalence, combining forces and moments, and system reduction are indispensable skills for civil engineers, enabling them to tackle the challenges of statics with confidence and precision.

Applications of Equivalence in Engineering Structures

In the realm of engineering structures, the application of equivalence principles plays a pivotal role in simplifying complex force systems, thereby facilitating a more streamlined and efficient analysis process. This approach is particularly beneficial in the design and evaluation of structures where multiple forces and moments are at play. By applying the concept of force-couple equivalence, engineers can transform a complicated system of forces into an equivalent force-couple system at a specific point, significantly reducing the complexity of static analysis.

The practical applications of this methodology are vast and varied across different types of engineering structures, from simple beams and frames to complex trusses and bridges. For instance, in the analysis of a beam subjected to various distributed and point loads, the engineer can calculate the resultant force and its location along the beam. This resultant force, combined with the resultant moment calculated about any point, often the support, can simplify the system to a single force and a couple moment. This simplification allows for the straightforward application of equilibrium equations, thereby enabling the determination of support reactions or internal forces with greater ease.

Moreover, in the context of truss structures, which are composed of numerous members connected at joints, the principle of equivalence is invaluable. By reducing the complex system of forces acting on a truss into a simpler model, engineers can more readily analyze the forces in individual truss members. This is achieved by first determining the overall resultant force and moment acting on the truss and then applying these simplified models to each node or joint, thus breaking down the problem into manageable parts.

Another significant application of equivalence principles is found in the design of retaining walls and foundations. In these cases, the earth pressures and other loads acting on the structure can be represented by equivalent force systems, simplifying the calculation of moments and forces necessary for stability analysis. This simplification is crucial for ensuring that the design can adequately withstand the imposed loads, with a clear understanding of the force distribution throughout the structure.

The process of simplifying complex force systems through the application of equivalence principles not only aids in the structural analysis but also enhances the efficiency of the design process. By reducing a system to its equivalent force and moment, engineers can more easily identify critical points, calculate necessary dimensions, and select appropriate materials, all while ensuring that the structure meets safety and performance criteria.

Furthermore, the application of these principles extends beyond the initial design phase, offering benefits in the evaluation of existing structures. Through the use of equivalence, engineers can assess the impact of modifications, such as the addition of new loads or alterations to the structural configuration, by comparing the original and modified systems in their simplified forms. This approach allows for a rapid assessment of the changes, facilitating timely and informed decision-making regarding necessary reinforcements or adjustments.

In conclusion, the applications of equivalence in engineering structures embody a fundamental aspect of statics that significantly contributes to the simplification and efficiency of structural analysis and design. Through the adept application of these principles, engineers are equipped to tackle the complexities inherent in diverse structural systems, ensuring both the integrity and functionality of the built environment.

Equilibrium of Rigid Bodies

Equilibrium Conditions

In the study of statics, particularly within the context of equilibrium of rigid bodies, the concept of **equilibrium conditions** plays a pivotal role. These conditions are foundational for analyzing and solving problems related to structures or components at rest or moving at a constant velocity. The primary tools used in this analysis are **free-body diagrams (FBDs)** and the **equations of equilibrium**. Understanding how to effectively apply these tools is essential for civil engineers aiming to ensure the stability and integrity of their designs.

Free-body diagrams are simplified representations of a system, isolated from its surroundings, with all applied forces, moments, and reactions depicted. Creating an accurate FBD is the first step in analyzing the equilibrium of a body. It involves identifying all forces acting on the body, including gravitational forces, applied loads, and support reactions. Each force is represented by a vector, indicating both its magnitude and direction. By isolating the body and illustrating these forces, engineers can systematically approach the problem, focusing on the interactions that affect equilibrium.

To analyze the equilibrium of a rigid body, we apply the **equations of equilibrium**, which are derived from Newton's first law of motion. For a body to be in equilibrium, the net force and the net moment acting on it must be zero. This condition is expressed mathematically by the following equations:

1. $\sum F_x = 0$ - The sum of all horizontal forces acting on the body must equal zero.
2. $\sum F_y = 0$ - The sum of all vertical forces acting on the body must equal zero.
3. $\sum M_O = 0$ - The sum of all moments about any arbitrary point O must equal zero.

These equations are the cornerstone of statics, allowing engineers to solve for unknown forces and moments acting on a body. When applying these equations, it's crucial to maintain consistency in the choice of coordinate direction (positive and negative axes) and the direction of moments (clockwise or counterclockwise).

In practice, solving for unknown forces and moments involves a systematic approach:

- **Step 1:** Draw a detailed free-body diagram, labeling all known and unknown forces and moments.

- **Step 2:** Choose a convenient coordinate system and resolve all forces into their components along these axes.
- **Step 3:** Apply the equations of equilibrium ($\sum F_x = 0$, $\sum F_y = 0$, and $\sum M_O = 0$) to the free-body diagram. This may involve summing forces in the horizontal and vertical directions and taking moments about a point that simplifies the calculation, ideally a point where the moment arm of unknown forces is zero.
- **Step 4:** Solve the resulting system of equations for the unknown quantities. This step may require the use of algebraic manipulation or matrix operations, depending on the complexity of the system.

The ability to accurately model a physical system with a free-body diagram and apply the equations of equilibrium is fundamental for civil engineers. This skill enables the analysis and design of structures and mechanical systems that are safe, efficient, and compliant with regulatory standards. Mastery of these concepts is not only crucial for passing the FE Civil Exam but also forms the basis of everyday engineering practice, ensuring that structures can withstand the forces and moments to which they are subjected.

Applications of Equilibrium

In the realm of statics, particularly when addressing the equilibrium of rigid bodies, practical applications serve as the bridge between theoretical concepts and real-world engineering challenges. The analysis of beams, frames, and multi-force members under equilibrium conditions exemplifies the direct application of static principles to ensure structural integrity and functionality. This section delves into these applications, highlighting the critical role of equilibrium in the design and analysis of various structural elements.

Beams, fundamental components in many engineering structures, are subjected to a variety of loads and reactions that must be meticulously analyzed to prevent failure. When a beam is in equilibrium, the sum of forces and moments acting on it equals zero. This condition is crucial for determining the beam's support reactions, which are essential for designing safe and reliable structures. For instance, consider a simply supported beam subjected to a uniform distributed load w over its entire length L. The reactions at the supports can be determined by applying the equations of equilibrium:

$$\sum F_y = 0 \Rightarrow R_A + R_B - wL = 0$$

$$\sum M_A = 0 \Rightarrow R_B \cdot L - \frac{wL^2}{2} = 0$$

where R_A and R_B are the reactions at the supports. Through these equations, one can solve for R_A and R_B, ensuring the beam's equilibrium and enabling further analysis, such as bending moment and shear force calculations.

Frames, which are structures composed of multiple members connected together, present a more complex scenario for equilibrium analysis due to the additional moments and forces at the connections. The equilibrium of a frame is analyzed by breaking it down into its constituent members and applying the conditions of equilibrium to each member individually. This process often involves considering both the external loads applied to the frame and the internal forces and moments at the connections. For example, in a rectangular frame with a horizontal load applied at one corner, each member must be analyzed separately to determine the internal forces and moments. The equilibrium equations for each member will include the contributions from the applied loads as well as the reactions and moments at the connections, ensuring the overall stability of the frame.

Multi-force members, characterized by the presence of several forces acting at different points along the member, require a detailed equilibrium analysis to determine the internal forces and moments. These members are crucial in trusses, where the axial forces in each member contribute to the overall stability of the structure. By applying the equilibrium conditions to each node or joint in the truss, engineers can solve for the unknown forces in the members. This analysis typically involves setting the sum of horizontal and vertical forces at each joint to zero:

$$\sum F_x = 0, \quad \sum F_y = 0$$

Through systematic application of these conditions across all joints, the axial forces in the truss members can be determined, ensuring the structure's equilibrium and facilitating the design of safe and efficient truss systems.

The practical applications of equilibrium in beams, frames, and multi-force members underscore the importance of a thorough understanding of statics principles for civil engineers. By applying these principles to analyze and design structural elements, engineers ensure that the structures can withstand the imposed loads without exceeding their strength limits, thereby safeguarding the well-being of the public and the longevity of the built environment. Through meticulous analysis and adherence to equilibrium conditions, the challenges posed by complex structural systems can be effectively addressed, highlighting the indispensable role of statics in the field of civil engineering.

Frames and Trusses

Analysis of Trusses: Joints and Sections Methods

The **analysis of trusses** is a fundamental aspect of statics, focusing on determining the forces in truss members, which are crucial for the design and safety of structures. Trusses are composed of slender members joined together at their ends, typically subjected to axial tension or compression. The methods of joints and sections are two primary approaches used in the analysis of trusses, allowing engineers to solve for unknown forces systematically.

Method of Joints begins with the assumption that all members of the truss are in equilibrium. This method involves isolating each joint within the truss and applying the conditions of equilibrium to solve for the unknown forces. For a joint to be in equilibrium, the sum of forces in both the horizontal and vertical directions must be zero. Mathematically, this can be expressed as:

$$\sum F_x = 0, \quad \sum F_y = 0$$

By applying these conditions to a free-body diagram of each joint, the forces acting on each member can be determined. It is essential to consider the direction of each force, where tension is considered positive and compression negative. This method is particularly effective for trusses with a simple configuration and a small number of members.

Method of Sections is a powerful technique used to analyze larger trusses or when the forces in a specific portion of the truss are required. This method involves cutting the truss into sections and analyzing one section at a time. By applying the equations of equilibrium to the section, the forces in the cut members can be solved directly. The section must be chosen such that no more than three unknown forces are cut, as there are only three equations of equilibrium available for solving these forces:

$$\sum F_x = 0, \quad \sum F_y = 0, \quad \sum M = 0$$

The moment equation, $\sum M = 0$, is particularly useful in the method of sections as it allows for the direct calculation of forces without the need for solving a system of equations. By taking moments about a point where two of the unknown forces intersect, the third force can be found directly. This approach significantly reduces the complexity of calculations, especially in trusses with a large number of members.

When applying both methods, it is crucial to accurately draw the free-body diagram of the joint or section under consideration, clearly indicating all known and unknown forces, as well as their directions. The choice between the method of joints and the method of sections often depends on the specific requirements of the problem and the complexity of the truss.

In practice, the analysis of trusses using these methods requires a systematic approach, starting from the supports or sections with the least number of unknowns. For efficient analysis, it is advisable to solve for support reactions first, using the overall equilibrium of the truss. This provides the necessary boundary conditions to proceed with the method of joints or sections.

The accuracy of truss analysis is paramount, as it directly impacts the design and safety of structural systems. Engineers must ensure that all assumptions made during the analysis, such as pinned connections and members only subjected to axial forces, accurately reflect the actual conditions of the truss. This meticulous approach to truss analysis ensures the reliability and integrity of engineering structures, aligning with the goal-oriented and analytical nature of the engineering profession.

Frame Analysis: Force and Moment Distribution

In the realm of statics, **frame analysis** is a pivotal concept that involves the examination of structures composed of multiple members connected at their ends to form a rigid structure. These frames are subjected to various loads and support conditions, necessitating a comprehensive understanding of how forces and moments are distributed throughout the structure. This section delves into the methodologies employed in analyzing frames with multiple members, focusing on the identification and calculation of force/moment distribution.

Frames, by their nature, are structures that support loads while maintaining their shape. The complexity of frame analysis arises from the need to consider both the geometry of the structure and the applied loads. The primary goal is to determine the internal forces (axial forces, shear forces, and bending moments) at various points within the members of the frame.

The analysis begins with the **free-body diagram (FBD)** of the entire frame, identifying all external forces and moments, including applied loads and reactions at supports. The equilibrium equations, based on Newton's First Law, are then applied to the entire frame to solve for unknown reactions. These equations are:

- $\sum F_x = 0$ (sum of horizontal forces)
- $\sum F_y = 0$ (sum of vertical forces)
- $\sum M = 0$ (sum of moments)

For frames that are statically determinate, these equilibrium equations are sufficient to solve for all unknowns. However, for statically indeterminate frames, additional compatibility conditions and methods, such as the method of superposition, are required.

Once the reactions are determined, the analysis proceeds to individual members of the frame. Each member is isolated, and a free-body diagram is drawn, considering the interactions at each connection or joint. It's crucial to accurately represent the direction and sense of the forces and moments at these joints, as they influence the internal force distribution within the members.

The **method of joints** and the **method of sections** are two fundamental approaches used in frame analysis:

- **Method of Joints:** This method involves isolating each joint within the frame and applying equilibrium equations to solve for unknown forces. It is particularly effective for determining the forces in members connected to a common joint.

- **Method of Sections:** This approach involves cutting through the frame to expose internal forces within members. By applying equilibrium equations to the section of the frame, one can solve for internal forces without analyzing the entire structure. This method is advantageous for finding forces in specific members without needing to solve for forces in the entire frame.

In analyzing the moment distribution, it's essential to consider the **bending moment diagrams** for each member. The bending moment at any point within a member is influenced by the type of loads applied (point loads, distributed loads) and the connection conditions at the ends of the member (fixed, pinned, roller). Calculating the bending moments involves integrating the shear force diagram or applying the moment-area theorems.

Centroid of Area

Centroid Basics

The centroid of an area, often referred to as the geometric center, is a critical concept in the field of statics and structural analysis, representing the point where an area's total shape balances evenly. This point is crucial for understanding various structural behaviors, as it is the location through which the resultant of gravitational forces acts in a body, assuming uniform density. The determination of the centroid is fundamental in the analysis and design of structures, as it affects the object's stability and stress distribution.

To mathematically locate the centroid of simple shapes, one must integrate over the shape's area to find the average position of all the infinitesimal elements that make up the area. The coordinates $(\bar{x}, \bar{y})$ of the centroid in a Cartesian coordinate system can be found using the formulas:

$$\bar{x} = \frac{1}{A}\int x\, dA$$

$$\bar{y} = \frac{1}{A}\int y\, dA$$

where A is the area of the shape, and x and y are the coordinates of an infinitesimal element dA of the area. These integrals average the positions of all points in the area, weighted by their distance from the reference axis.

For simple geometric shapes, such as rectangles, triangles, circles, and semicircles, the centroids can be easily determined due to their symmetry. The principle of symmetry simplifies the process, as it dictates that the centroid lies along the axes of symmetry. For instance, the centroid of a rectangle is located at its geometric center because it has two axes of symmetry that intersect at the center. Similarly, for a triangle, the centroid is found at the intersection of its medians, which is one-third of the distance from each vertex to the midpoint of the opposite side.

In the case of composite areas, which are shapes made up of several simple geometric figures, the centroid can be found by decomposing the composite area into its constituent simple shapes, finding the centroid of each simple shape, and then calculating the weighted average of these centroids. The weights in this average are proportional to the area of each simple shape. This method leverages the additive property of integrals, allowing for the centroid of a complex shape to be determined by summing the moments of its parts about a reference axis and dividing by the total area. The formula for the centroid of a composite area is given by:

$$\bar{x} = \frac{\sum(A_i \bar{x}_i)}{\sum A_i}$$

$$\bar{y} = \frac{\sum(A_i \bar{y}_i)}{\sum A_i}$$

where A_i is the area of the ith simple shape, and $\bar{x}_i$ and $\bar{y}_i$ are the centroid coordinates of the ith simple shape. This approach is particularly useful in engineering applications where complex shapes are prevalent, and precise knowledge of the centroid is necessary for analysis and design purposes.

Understanding the concept of centroids and mastering the technique to locate them in various shapes are indispensable skills for engineers. This knowledge is not only fundamental in statics but also in dynamics, strength of materials, and structural analysis, where the distribution of forces and moments is influenced by the position of the centroid. The ability to accurately determine centroids ensures that engineers can predict the behavior of structures under loads, design efficient structural members, and ensure the safety and reliability of their projects.

Composite Areas: Centroids of Shapes

Determining the centroids of composite areas is a fundamental task in the field of statics, essential for analyzing the structural integrity and designing engineering projects. Composite areas are typically formed by combining simple geometric shapes such as rectangles, triangles, circles, and semicircles, whose individual centroids are known or can be easily calculated. The challenge arises when these simple shapes are configured in a manner that creates a complex shape, whose centroid is not immediately apparent. The process of finding the centroid of such composite areas involves breaking down the complex shape into its constituent simple shapes, calculating the centroid of each, and then using these centroids to find the centroid of the entire composite area.

The decomposition method leverages the principle that the centroid of a composite area can be determined by summing the moments of its constituent areas about a reference axis and dividing by the total area. This method is mathematically represented by the equations:

$$\bar{x} = \frac{\sum(A_i \bar{x}_i)}{\sum A_i}$$

$$\bar{y} = \frac{\sum(A_i \bar{y}_i)}{\sum A_i}$$

where A_i represents the area of the ith simple shape, $\bar{x}_i$ and $\bar{y}_i$ are the x and y coordinates of the centroid of the ith shape, respectively. The summation extends over all constituent shapes of the composite area. This approach effectively treats each component shape as a point mass at its centroid, contributing to the overall centroid location of the composite shape based on its area.

Integral methods provide another avenue for finding centroids, particularly useful when dealing with irregular shapes or when the composite area cannot be easily broken down into simpler shapes. This method involves setting up an integral that sums the moments of infinitesimally small elements (dA) of the area about a reference axis. The coordinates of the centroid $(\bar{x}, \bar{y})$ are found by integrating over the entire area of the shape. The formulas for these coordinates are given by:

$$\bar{x} = \frac{1}{A} \int x \, dA$$

$$\bar{y} = \frac{1}{A} \int y \, dA$$

where A is the total area of the composite shape, and x and y are the coordinates of an element dA with respect to the chosen reference axes. The integral method is particularly advantageous when the geometry of the composite area is defined by functions, allowing for the precise calculation of the centroid through mathematical integration.

In applying these methods, it is crucial to choose a suitable reference axis or axes, which can simplify calculations. For symmetric shapes, aligning the reference axis with the axis of symmetry can significantly reduce the complexity of the problem. When dealing with composite areas, attention must be paid to the sign of the coordinates $\bar{x}_i$ and $\bar{y}_i$, especially if parts of the composite area lie in different quadrants of the Cartesian coordinate system.

The accuracy of the centroid location is paramount, as it affects the analysis of structural behavior under loads, the design of structural members, and the prediction of stress distribution and deformation. Engineers must adeptly apply these methods to ensure the reliability and safety of structures, taking into consideration the nuances of each approach and the specific characteristics of the composite area under examination. The ability to accurately decompose complex shapes into simpler components and apply integral methods for irregular geometries is a testament to the engineer's skill in leveraging mathematical principles to solve practical problems in statics and structural analysis.

Area Moments of Inertia

Basic Concepts: Moment of Inertia and Radius of Gyration

The **moment of inertia** (I) is a fundamental concept in the field of statics and structural analysis, quantifying an object's resistance to angular acceleration about a specific axis. It is integral to understanding how different shapes and materials will behave under various loading conditions, particularly in bending and torsional scenarios. The moment of inertia is dependent not only on the mass of an object but also on how that mass is distributed relative to the axis of rotation. For engineering applications, the moment of inertia is crucial in the design and analysis of beams, columns, and other structural elements to ensure they can withstand applied loads without excessive deformation or failure.

Mathematically, the moment of inertia for a discrete system is expressed as $I = \sum m_i r_i^2$, where m_i represents the mass of each particle in the body, and r_i is the distance of each particle from the axis of rotation. For continuous bodies, the moment of inertia is determined by integrating the differential mass (dm) over the entire volume of the body, with the formula $I = \int r^2 dm$, where r is the distance from the axis of rotation to the differential element of mass (dm).

The **radius of gyration** (k) is another pivotal concept, providing a simplified measure of how a body's area or mass is distributed about an axis. It is defined as the square root of the ratio of the moment of inertia to the area (A) or mass (m) of the body, depending on whether the context is area moments of inertia or mass moments of inertia, respectively. The formula for the radius of gyration in terms of area is given by $k = \sqrt{\frac{I}{A}}$, and in terms of mass by $k = \sqrt{\frac{I}{m}}$. The radius of gyration offers a single value that characterizes the distribution of the area or mass about the axis, simplifying the analysis and comparison of different structural elements.

In structural engineering, the moment of inertia plays a critical role in determining the bending stress (σ) in a beam, which is calculated using the formula $\sigma = \frac{My}{I}$, where M is the moment at a given section of the beam, y is the distance from the neutral axis to the point where the stress is being calculated, and I is the moment of inertia about the neutral axis. This relationship highlights the importance of the moment of inertia in designing beams and other structural components to resist bending.

Furthermore, the moment of inertia is essential in calculating the deflection (δ) of beams under load, with the deflection inversely proportional to the moment of inertia, as shown in the formula $\delta = \frac{FL^3}{3EI}$ for a simply supported beam with a central point load (F), where L is the length of the beam and E is the modulus of elasticity of the material. This inverse relationship underscores the significance of maximizing the moment of inertia to minimize deflection and ensure structural integrity.

The radius of gyration is particularly useful in the analysis and design of columns, as it is directly related to the slenderness ratio (λ), which is a key factor in determining the buckling load of a column. The slenderness ratio is defined as the ratio of the effective length (L) of the column to its radius of gyration (k), with the formula $\lambda = \frac{L}{k}$. A higher slenderness ratio indicates a greater likelihood of buckling, guiding engineers in selecting appropriate dimensions and materials to prevent instability.

The moment of inertia and radius of gyration are indispensable tools in the engineer's toolkit, enabling the precise analysis and design of structural elements to meet safety and performance criteria. Their application spans a wide range of engineering problems, from determining the stress and deflection of beams to assessing the stability of columns against buckling, illustrating their broad relevance and importance in the field of structural engineering.

Composite Shapes: Moments of Inertia Calculation

Calculating the moments of inertia for composite shapes is a critical skill for engineers, particularly when analyzing the structural integrity of beams, columns, and other elements subjected to bending and torsional stresses. The moment of inertia, I, quantifies an object's resistance to angular acceleration about a specific axis, and its calculation for composite shapes requires a methodical approach that combines the moments of inertia of simpler, constituent shapes using the addition and parallel axis theorem.

For a composite area made up of simpler shapes, the total moment of inertia about a given axis can be found by summing the moments of inertia of each individual shape about the same axis. This process is straightforward when the axis of interest coincides with an axis of symmetry for each of the constituent shapes. However, when the individual shapes' axes of symmetry do not align with the desired axis of the composite shape, the parallel axis theorem must be applied to calculate the moment of inertia of each shape about the common axis.

The parallel axis theorem states that the moment of inertia of a shape about any axis parallel to an axis through its centroid is equal to the moment of inertia of the shape about its centroidal axis plus the product of the area of the shape and the square of the distance between the two axes. Mathematically, this is expressed as:

$$I = I_c + Ad^2$$

where I is the moment of inertia about the parallel axis, I_c is the moment of inertia about the centroidal axis, A is the area of the shape, and d is the distance between the centroidal axis and the parallel axis.

To calculate the moment of inertia of a composite shape, one must first identify and separate the composite shape into its simpler constituent shapes. The moment of inertia of each of these simpler shapes about their own centroidal axes is either known from standard tables or can be calculated using basic formulas. Next, if the axis about which the total moment of inertia of the composite shape is desired does not pass through the centroids of the constituent shapes, the parallel axis theorem is applied to find the moment of inertia of each shape about the common axis.

The steps involve calculating the distance d from the centroid of each constituent shape to the common axis and then applying the parallel axis theorem to find each shape's moment of inertia about the common axis. These individual moments of inertia are then summed to obtain the total moment of inertia of the composite shape:

$$I_{total} = \sum (I_c + Ad^2)$$

This equation underscores the additive nature of moments of inertia and the utility of the parallel axis theorem in simplifying the calculation for composite shapes. It is essential for engineers to accurately perform these calculations, as the moment of inertia plays a crucial role in predicting the behavior of structural elements under load, determining their bending stress, deflection, and overall stability.

In practice, the calculation of moments of inertia for composite shapes enables engineers to design more efficient and safer structures by optimizing the distribution of material and cross-sectional geometry to resist applied loads. This process, while mathematically rigorous, provides a foundation for the structural analysis and design tasks that are central to civil engineering projects, ensuring that structures can withstand the forces and moments they encounter during their service life.

Static Friction

Friction Basics

Friction plays a pivotal role in the realm of statics, influencing the behavior of objects at rest and in motion. Understanding the fundamentals of frictional forces, along with the coefficients of static and kinetic friction, is essential for engineers to design and analyze various structures and mechanical systems effectively.

Frictional forces arise due to the interaction between the surfaces of two bodies in contact. These forces oppose the relative motion or the tendency of motion between the surfaces. The magnitude of the frictional force (F_f) can be calculated using the formula $F_f = \mu N$, where μ represents the coefficient of friction, and N is the normal force exerted by the surface on the object.

The coefficient of friction is a dimensionless scalar value that describes the ratio of the force of friction between two bodies and the force pressing them together. It varies depending on the materials of the surfaces in contact and their condition (e.g., dry, lubricated, rough, smooth). There are two primary types of friction coefficients: static (μ_s) and kinetic (μ_k).

Static friction acts on objects when they are at rest relative to each other. It must be overcome to initiate motion. The coefficient of static friction (μ_s) quantifies the frictional force that needs to be overcome to start moving an object. It is generally higher than the coefficient of kinetic friction for the same set of materials, indicating that more force is required to initiate movement than to maintain it. The maximum static frictional force ($F_{s_{max}}$) that can be applied before motion begins is given by $F_{s_{max}} = \mu_s N$.

Kinetic friction, on the other hand, acts on objects that are already in motion relative to each other. The coefficient of kinetic friction (μ_k) is used to calculate the frictional force opposing the motion once it has started. This force is given by $F_k = \mu_k N$. Kinetic friction is usually less than static friction, which explains why it is easier to keep an object moving than to start moving it from rest.

The distinction between static and kinetic friction is crucial for solving problems in statics and dynamics. For instance, when analyzing the motion of a block on an inclined plane, the type of frictional force acting on the block changes from static to kinetic at the instant the block starts to slide. Engineers must account for these forces accurately to predict the behavior of systems under various loads and conditions.

The concepts of frictional forces and coefficients of friction are fundamental in the study of statics, providing a basis for understanding the resistance to motion between contacting surfaces. Mastery of these concepts enables engineers to design safer and more efficient structures and mechanical systems by accurately predicting and controlling the effects of friction.

Friction in Inclined Planes and Wedges

In the realm of static friction applications, particularly in engineering mechanics, the analysis of objects on inclined planes and the use of wedges are pivotal for understanding how frictional forces influence motion and stability. These scenarios are common in various engineering problems and designs, from simple machines to complex structural components. The mathematical treatment of these applications not only solidifies the understanding of static friction but also equips engineers with the tools to predict and manipulate the behavior of physical systems under frictional influence.

When considering an object resting on an inclined plane, the gravitational force acting on the object can be decomposed into two components: one perpendicular to the plane, and one parallel to it. The normal force (N), which is equal in magnitude and opposite in direction to the perpendicular component of gravity, plays a crucial role in determining the frictional force (F_f) acting on the object. The static frictional force can be expressed as $F_f = \mu_s N$, where μ_s is the coefficient of static friction between the object and the plane. This relationship is critical in determining whether an object will remain stationary or start sliding down the plane. The angle of the incline, θ, directly influences both components of the gravitational force and, consequently, the frictional force. The condition for impending motion, where the object is on the verge of sliding, is reached when the parallel component of the gravitational force equals the maximum static frictional force. Mathematically, this is represented as $mg\sin(\theta) = \mu_s mg\cos(\theta)$, where m is the mass of the object and g is the acceleration due to gravity. Solving for θ gives the critical angle at which the object will start to slide.

Wedges, another fundamental application of static friction, are utilized to convert a force applied along the length of the wedge into a force normal to the surface being acted upon. The analysis of wedges involves determining the frictional forces at the interfaces between the wedge and the surface or object it is moving or splitting. Similar to inclined planes, the equilibrium conditions for a wedge require that the sum of the forces in any direction and the sum of the moments about any point must be zero. The effectiveness of a wedge, or its mechanical advantage, is significantly influenced by the coefficient of static friction. A higher coefficient of static friction means that a smaller force is required to either lift an object or hold it in place with the wedge. The equations governing the forces in wedge applications are derived from the principles of static equilibrium and frictional forces, where the normal and frictional forces at the contact surfaces dictate the required input force to achieve a desired outcome.

Impending motion problems in both inclined planes and wedges are essential for engineers to understand and predict the behavior of systems where friction plays a critical role. These problems often involve determining the conditions under which motion is about to occur, which requires a thorough understanding of the principles of static friction and the ability to apply these principles to real-world scenarios. The analysis of such problems not only aids in the design and optimization of mechanical systems and structures but also in the development of safety protocols and measures to prevent unwanted motion or slippage.

The study of static friction in the context of inclined planes, wedges, and impending motion provides a comprehensive understanding of how frictional forces can be harnessed or counteracted in engineering applications. Through meticulous analysis and application of fundamental principles, engineers can design more efficient, safe, and reliable systems that account for the effects of static friction.

Chapter 5: Dynamics

Kinematics: Motion of Particles and Rigid Bodies

Kinematics is the branch of mechanics that deals with the motion of objects without considering the forces that cause the motion. In the context of the FE Civil Exam, understanding kinematics is crucial for analyzing and solving problems related to particles and rigid bodies. This section delves into the motion of particles along rectilinear and curvilinear paths, velocity, acceleration, rigid body rotation, angular velocity, and relative motion.

Motion of Particles: Particles move in space and can follow either rectilinear (straight-line) paths or curvilinear (curved) paths. The position of a particle is defined relative to a chosen reference point and is described by its coordinates in a given coordinate system.

- **Rectilinear Motion**: When a particle moves along a straight path, its motion is said to be rectilinear. The position of the particle can be described by a single coordinate x, y, or z, depending on the direction of motion. The velocity (v) of the particle is the rate of change of its position with respect to time and is given by $v = \dfrac{dx}{dt}$ for motion along the x-axis. Acceleration (a) is the rate of change of velocity with respect to time, given by $a = \dfrac{dv}{dt}$.

- **Curvilinear Motion**: For motion along a curved path, the position of the particle is described by two or three coordinates, such as x, y (in a plane), or x, y, z (in space). The velocity of the particle is still the rate of change of its position, but it now has components in each direction of the coordinate system. The acceleration of a particle in curvilinear motion has two components: tangential acceleration, which is the rate of change of speed along the path, and normal or centripetal acceleration, which is directed towards the center of curvature of the path.

Rigid Body Rotation: A rigid body is an object that does not deform under the action of forces. In rigid body rotation, all points in the body move in circular paths around a common axis of rotation, and the body maintains its shape.

- **Angular Velocity** (ω): This is the rate of change of the angular displacement of the rigid body and is measured in radians per second. It is given by $\omega = \frac{d\theta}{dt}$, where θ is the angular displacement.

- **Angular Acceleration** (α): This is the rate of change of angular velocity with respect to time, given by $\alpha = \frac{d\omega}{dt}$.

Relative Motion: The motion of a particle or rigid body can be described relative to a moving reference frame as well as to a stationary one. The velocity and acceleration of an object as observed from a moving reference frame are different from those observed from a stationary frame. The relative velocity of an object is the velocity of the object in the stationary frame minus the velocity of the moving frame. Similarly, the relative acceleration is the acceleration of the object in the stationary frame minus the acceleration of the moving frame.

Mass Moments of Inertia

The **mass moment of inertia** is a critical concept in dynamics, representing the rotational inertia of a body. It quantifies how the distribution of mass affects an object's resistance to angular acceleration around an axis. For engineers preparing for the FE Civil Exam, understanding how to calculate the mass moment of inertia for both simple and composite bodies, as well as applying the parallel axis theorem, is essential for solving dynamics problems effectively.

For a simple body, the mass moment of inertia (I) about an axis can be calculated using the integral $I = \int r^2 dm$, where r is the distance from the axis of rotation to the element of mass (dm). This formula is the basis for deriving expressions for common geometric shapes. For example, the mass moment of inertia of a solid cylinder about its central axis is given by $I = \frac{1}{2}MR^2$, where M is the mass of the cylinder and R is its radius.

When dealing with composite bodies, the total mass moment of inertia is the sum of the individual moments of inertia of its components. Each component's moment of inertia is calculated about the same axis of rotation and then summed. This approach simplifies the calculation for complex structures by breaking them down into basic shapes whose moments of inertia are well-known or easier to compute.

The **parallel axis theorem** is a powerful tool in dynamics, allowing the calculation of an object's moment of inertia about any axis, given its moment of inertia about a parallel axis through the object's center of mass and the distance between the two axes. The theorem is expressed as $I = I_{CM} + Md^2$, where I is the moment of inertia about the new axis, I_{CM} is the moment of inertia about the center of mass axis, M is the total mass of the body, and d is the distance between the two axes. This theorem is particularly useful for engineering applications where the axis of rotation does not pass through the center of mass of the body.

In dynamics, the mass moment of inertia plays a crucial role in analyzing the rotational motion of bodies. It affects the angular velocity, angular acceleration, and the dynamic response of structures and mechanical systems to applied torques. For instance, in the design of flywheels, gears, and rotating machinery, engineers must accurately calculate the mass moment of inertia to ensure stability and efficient operation. Similarly, in the analysis of structural elements subjected to dynamic loads, such as seismic forces, the distribution of mass and the corresponding moments of inertia significantly influence the behavior of the structure.

The concepts of mass moment of inertia and the application of the parallel axis theorem provide engineers with essential tools for addressing a variety of dynamics problems, including the motion of particles and rigid bodies, as well as the design and analysis of mechanical systems and structures. Proficiency in these concepts is vital for success on the FE Civil Exam and serves as a fundamental basis for further studies and professional practice in engineering dynamics.

Force-Acceleration and Rigid Body Dynamics

Newton's second law of motion is foundational in understanding the dynamics of particles and rigid bodies, particularly when analyzing force-acceleration relationships. This law states that the force acting on an object is equal to the mass of the object multiplied by its acceleration ($F = ma$). This principle is crucial for solving problems related to linear and angular motion in the FE Civil Exam.

Linear Motion: In the context of linear motion, Newton's second law can be applied to determine the acceleration of a particle given the net force acting on it and its mass. For a particle of mass m subjected to a net force F, the acceleration a can be calculated as $a = \frac{F}{m}$. This relationship is pivotal in engineering applications, such as determining the forces in structural elements or the acceleration of vehicles.

Angular Motion: When considering angular motion, Newton's second law is extended to rotational systems. The torque (τ) acting on a body is equal to the moment of inertia (I) of the body multiplied by its angular acceleration (α), expressed as $\tau = I\alpha$. The moment of inertia is a measure of an object's resistance to changes in its rotational motion, analogous to mass in linear motion. This equation is essential for analyzing the dynamics of rotating systems, such as gears and wheels.

Free-Body Diagrams (FBDs): Free-body diagrams are invaluable tools for visualizing the forces acting on an object and for applying Newton's second law. An FBD represents all the external forces and moments acting on a body, simplifying the analysis of mechanical systems. By drawing an FBD, engineers can systematically apply Newton's second law to solve for unknown forces, moments, and resulting accelerations.

Rigid Body Dynamics: The study of rigid body dynamics involves analyzing the motion of bodies in which deformation is negligible. In rigid body dynamics, the equations of motion are derived by applying Newton's second law to the system's center of mass for translational motion and about the center of mass for rotational motion. The combined translational and rotational dynamics are described by the equations $F = ma$ for the center of mass and $\tau = I\alpha$ for rotation about the center of mass, where F and τ are the net external force and torque, respectively.

The analysis of rigid body dynamics is further complicated by the need to consider both the linear and angular components of motion. For instance, when a force is applied off-center on a rigid body, it can cause both translation and rotation. Engineers must decompose forces into components acting through the center of mass (causing translation) and components causing rotation (torques) to accurately predict the body's motion.

In engineering practice, understanding force-acceleration relationships and rigid body dynamics is essential for designing and analyzing mechanical systems, structures, and components subjected to various forces and moments. Mastery of these concepts, supported by the skillful use of free-body diagrams, enables engineers to solve complex dynamics problems, ensuring the safety, reliability, and efficiency of engineering solutions.

Work, Energy, and Power Principles

The principles of **work**, **energy**, and **power** are foundational in the study of dynamics, providing engineers with the tools to analyze and design systems under various conditions. **Work** is defined as the product of force and displacement in the direction of the force and is given by the formula $W = F \cdot d \cdot \cos(\theta)$, where F is the force applied, d is the displacement of the object, and θ is the angle between the force and the displacement vector. This concept is crucial in understanding how forces cause objects to move and how energy is transferred in the process.

Energy represents the capacity to do work. It exists in various forms, among which **kinetic** and **potential** energies are most relevant to dynamics. **Kinetic energy** (KE) is the energy of motion, calculated by the equation $KE = \frac{1}{2}mv^2$, where m is the mass of the object and v is its velocity. **Potential energy** (PE), on the other hand, is the stored energy of position. For objects near Earth's surface, gravitational potential energy can be calculated with $PE = mgh$, where g is the acceleration due to gravity, and h is the height above a reference point.

The **conservation of energy principle** states that energy cannot be created or destroyed, only transformed from one form to another. This principle is a powerful tool in engineering, allowing for the analysis of systems where energy is transferred between kinetic and potential forms without loss. For example, in a pendulum, energy oscillates between kinetic and potential, with total energy conserved if air resistance and friction are negligible.

Power is the rate at which work is done or energy is transferred, quantified as $P = \frac{W}{t}$, where W is work and t is the time over which the work is performed. In dynamic systems, understanding power is essential for designing engines, motors, and other machinery where efficiency and energy output are critical.

In the context of **rigid body applications**, these principles help in analyzing the motion and stability of structures and mechanical systems. For instance, in calculating the work done by torques in rotating systems, the concept of work is extended to $W = \tau\theta$, where τ is the torque and θ is the angular displacement. This extension is vital for understanding how forces and energy interact in systems involving rotational motion.

The **work-energy theorem**, which states that the work done on an object is equal to the change in its kinetic energy ($W = \Delta KE$), provides a method to analyze the motion of particles and rigid bodies without directly solving the forces involved. This theorem simplifies the calculation of velocities and displacements in systems where multiple forces act, such as in vehicle crash analysis or the design of roller coasters.

In engineering practice, applying these principles allows for the efficient design and analysis of systems ranging from simple machines to complex structures subjected to dynamic forces. By understanding how work, energy, and power interact, engineers can predict system behavior under various conditions, optimize performance, and ensure safety and reliability. The ability to analyze energy transformations and power requirements is indispensable in the development of sustainable and innovative engineering solutions, reflecting the role of engineers in addressing global challenges through thoughtful design and responsible resource management.

Chapter 6: Mechanics of Materials

Shear and Moment Diagrams

Diagram Basics: Shear and Moment Diagrams

Understanding the interactions between loads, shear forces, and bending moments is essential for analyzing beams, which serve as vital components in various structures. This section focuses on the fundamentals of how these elements relate to one another and how to depict them using shear and moment diagrams, which are crucial tools for engineers involved in the design and analysis of structural elements.

Loads applied to a beam produce reactions at the supports due to the beam's resistance to bending. These loads can be concentrated (point loads), distributed, varying, or moment loads. The type and distribution of the load affect the shear force and bending moment along the length of the beam.

Shear force at a section of a beam is a measure of the internal force parallel to the cross-section that is developed as the beam resists bending. It varies along the length of the beam as the applied load changes. The shear force between any two points on the beam is constant if the load is uniform. However, it changes at points where the load varies. The convention is to consider downward forces on the left of the section as positive, leading to a positive shear force that causes a clockwise rotation.

The **bending moment** at a section within a beam represents the internal moment that the section is subjected to due to the applied loads. It is a measure of the bending effect due to forces acting perpendicular to the beam's longitudinal axis. The bending moment at a given section is the sum of moments about that section. Moments causing concave curvature (sagging) are considered positive, while those causing convex curvature (hogging) are negative.

To **draw shear and moment diagrams**, one must first calculate the reactions at the supports using the conditions of static equilibrium: $\sum F = 0$ and $\sum M = 0$. After determining the support reactions, the next step is to calculate the shear force and bending moment at key points along the beam, such as at points of applied loads, points of load changes, and at the supports.

1. **Shear Force Diagram (SFD)**: Start from the left end of the beam and move towards the right, plotting the shear force value at each point. The shear force between two points is constant if there is no load, increases or decreases linearly with a uniformly distributed load (UDL), and changes abruptly at points of concentrated loads.

2. **Bending Moment Diagram (BMD)**: Similar to the SFD, the BMD is plotted by moving from one end of the beam to the other. The bending moment at any point is the area under the shear force diagram up to that point. The moment increases or decreases linearly under no load or under a uniformly distributed load, respectively, and has a parabolic variation under a UDL due to the linear change in shear.

For example, consider a simply supported beam with a point load P at the midpoint. The reactions at the supports are $P/2$. The shear force is $+P/2$ from the left support to the point of the load, where it drops to $-P/2$ and remains constant to the right support. The bending moment is zero at the supports, increases linearly to a maximum of $P \cdot L/4$ at the midpoint (where L is the length of the beam), and then decreases linearly back to zero at the right support.

Shear and moment diagrams are graphical representations that provide valuable information about the internal forces and moments within a beam, enabling engineers to design and analyze structural elements effectively.

Applications in Structural Design

In the realm of structural design, the ability to identify critical points and maximum/minimum values on shear and moment diagrams is paramount. These diagrams serve as a visual representation of the internal forces and moments experienced by a beam under various loading conditions. By accurately interpreting these diagrams, engineers can make informed decisions regarding the design and optimization of structural elements to ensure safety, efficiency, and cost-effectiveness.

Critical Points on shear and moment diagrams are locations where the value of shear force or bending moment changes direction or reaches a local maximum or minimum. These points are typically found at locations of applied loads, supports, and points of discontinuity in the loading pattern. Identifying these points is crucial for determining the design values for shear and bending moment, which directly influence the sizing of structural components and the selection of materials.

For instance, in a **Shear Force Diagram (SFD)**, critical points are identified where the shear force curve intersects the horizontal axis, indicating points of zero shear. These points are essential for understanding the regions in a beam where shear reinforcement may be necessary to prevent shear failure. Additionally, abrupt changes in the slope of the shear force diagram indicate the application of concentrated loads or moments, necessitating careful consideration in the design to accommodate these forces.

In a **Bending Moment Diagram (BMD)**, the maximum and minimum values represent the greatest positive and negative moments a beam will experience. The locations of these maxima and minima are critical for determining where the beam will experience the most significant bending stresses. Designing for these stresses involves selecting beam dimensions and materials that can safely resist the expected moments, thereby preventing failure modes such as excessive deflection or cracking.

For example, consider a simply supported beam subjected to a uniform distributed load. The maximum bending moment occurs at the midpoint of the beam and can be calculated using the formula $M_{max} = \dfrac{wL^2}{8}$, where w is the load per unit length and L is the span of the beam. This maximum moment value is a critical design parameter, dictating the necessary depth and reinforcement of the beam to ensure it can safely support the applied loads.

Furthermore, the use of **superposition** in analyzing complex loading conditions allows for the decomposition of complicated load patterns into simpler, more manageable components. By analyzing each component separately and then combining the results, engineers can more easily identify critical points and maximum/minimum values on the shear and moment diagrams. This method enhances the accuracy of the analysis and the efficiency of the design process.

Incorporating these analytical techniques into structural design ensures that all potential failure points are adequately addressed, leading to structures that are not only safe and reliable but also optimized for performance and cost. By diligently applying the principles of shear and moment diagram analysis, engineers contribute to the creation of resilient and sustainable built environments.

Stresses and Strains

Stress Analysis Fundamentals

In the realm of mechanics of materials, understanding the nuances of **stress analysis** is pivotal for engineering applications. Stress analysis encompasses the evaluation of **axial, torsional, bending,** and **shear stresses**, each of which plays a critical role in the integrity and functionality of structural components.

Axial stress (σ) arises when a force is applied directly along the longitudinal axis of a member, leading to either tension or compression. It is calculated as the force (F) divided by the cross-sectional area (A) of the member, expressed by the formula $\sigma = \frac{F}{A}$. In tension, axial stress tends to elongate the member, while in compression, it seeks to shorten it.

Torsional stress (τ) occurs when a member is subjected to a twisting action caused by a torque (T) applied about its longitudinal axis. This stress is most relevant in shafts and is given by $\tau = \frac{T \cdot r}{J}$, where r is the outer radius of the shaft and J is the polar moment of inertia. Torsional stress is critical in understanding the rotational behavior and failure modes of cylindrical members.

Bending stress (σ_b) is induced in a beam or a similar member when an external moment or transverse load causes it to bend, leading to tension on one side and compression on the other. The bending stress at any point in the beam's cross-section is calculated using $\sigma_b = \frac{M \cdot y}{I}$, where M is the moment at the section, y is the distance from the neutral axis, and I is the moment of inertia of the cross-section. This stress is a primary consideration in the design of beams and other flexural members.

Shear stress (τ) results from forces that act parallel to the cross-sectional area of the material, causing sliding failure along the plane of the material. For a rectangular cross-section, it is often approximated as $\tau = \dfrac{V \cdot Q}{I \cdot b}$, where V is the shear force, Q is the first moment of area, I is the moment of inertia, and b is the width of the section in the shear plane. Shear stress is a key factor in the design of fasteners and connections.

The relationship between stress and strain for materials is depicted through **stress-strain diagrams**, which provide a graphical representation of a material's response to stress. The slope of the elastic portion of the diagram is known as the **modulus of elasticity** (E), which measures the material's stiffness. The diagram also features the yield strength, ultimate strength, and the fracture point, offering insights into the material's ductility, resilience, and toughness.

Material properties significantly influence stress analysis. For instance, ductile materials, like most metals, exhibit a significant region of plastic deformation on their stress-strain diagram, allowing for energy absorption and redistribution of stress. Conversely, brittle materials, such as ceramics, have a minimal plastic region, leading to sudden failure upon reaching their ultimate strength.

In summary, a comprehensive understanding of axial, torsional, bending, and shear stresses, coupled with the insights provided by stress-strain diagrams, equips engineers with the necessary tools to predict the behavior of materials and structures under various loading conditions. This knowledge is crucial for ensuring the safety, reliability, and efficiency of engineering designs, addressing the analytical and goal-oriented nature of the engineering profession.

Thermal Stresses and Strain

Thermal stresses arise when temperature changes lead to the expansion or contraction of materials, which, in turn, can induce stress and strain within a structure. These stresses are particularly significant in engineering materials and structures because they can affect their integrity, performance, and durability. Understanding the fundamentals of thermal stress and strain is crucial for engineers to design structures that can withstand temperature variations without failure.

The basic principle behind thermal stress is the tendency of materials to change in dimension in response to temperature changes. This dimensional change is quantified by the coefficient of thermal expansion (α), which is a material property that describes how much a material expands per unit length per degree change in temperature. The thermal strain ($\epsilon_{thermal}$) induced by a temperature change (ΔT) can be expressed as:

$$\epsilon_{thermal} = \alpha \Delta T$$

When a material is constrained and cannot freely expand or contract in response to temperature changes, thermal stress ($\sigma_{thermal}$) is generated. The magnitude of this stress can be calculated using the formula:

$$\sigma_{thermal} = E \cdot \epsilon_{thermal} = E \cdot \alpha \Delta T$$

where E is the modulus of elasticity of the material. This equation highlights the direct relationship between thermal stress and both the coefficient of thermal expansion and the modulus of elasticity. Materials with a high coefficient of thermal expansion or a high modulus of elasticity will experience greater thermal stresses for a given temperature change.

Compatibility conditions refer to the constraints that are applied to a structure or component, which can significantly influence the development of thermal stresses. For example, if two materials with different coefficients of thermal expansion are bonded together and subjected to a temperature change, the material with the higher coefficient will want to expand more than the other. This difference in expansion can induce significant thermal stresses at the interface of the two materials, potentially leading to failure modes such as delamination or cracking.

In the context of structural engineering, it is essential to consider thermal stresses in the design phase to ensure the longevity and safety of structures. This consideration is particularly important for structures exposed to wide temperature variations or those constructed using materials with significantly different thermal expansion coefficients. Engineers can mitigate the effects of thermal stresses through various design strategies, such as using expansion joints in bridges and buildings to allow for free expansion and contraction of the structure without inducing undue stress.

Furthermore, the analysis of thermal stresses requires a thorough understanding of the thermal properties of materials, as well as the environmental conditions to which the structure will be exposed. Advanced computational tools, such as finite element analysis (FEA), are often employed to model the complex interactions between temperature changes, material properties, and structural constraints, enabling engineers to predict and address potential issues related to thermal stresses.

Thermal stresses and strains are critical considerations in the design and analysis of engineering structures and materials. By understanding the relationship between temperature changes, material properties, and structural constraints, engineers can develop strategies to mitigate the adverse effects of thermal stresses, ensuring the structural integrity and longevity of their designs.

Deformations

Axial and Torsional Deformations

Axial and torsional deformations are fundamental concepts in the mechanics of materials, crucial for understanding how forces and torques affect the structural integrity and performance of engineering components. Axial deformation occurs when a force is applied along the longitudinal axis of a material, causing it to elongate or shorten. This type of deformation is characterized by changes in the length of the material without a change in its cross-sectional area. The amount of axial deformation can be quantified using Hooke's Law for linearly elastic materials, which states that the deformation (δ) is directly proportional to the applied load (F) and the original length (L_0) of the material, and inversely proportional to the cross-sectional area (A) and the modulus of elasticity (E) of the material. The formula for axial deformation is given by:

$$\delta = \frac{F L_0}{A E}$$

Torsional deformation, on the other hand, occurs when a torque or twisting force is applied to a material, causing it to twist around its longitudinal axis. This type of deformation is particularly relevant for cylindrical shafts and rods subjected to torsional loads. The angle of twist (θ) in radians that a material undergoes due to a torsional load can be calculated using the formula:

$$\theta = \frac{TL}{JG}$$

where T is the applied torque, L is the length of the material, J is the polar moment of inertia of the cross-sectional area, and G is the shear modulus of the material. The polar moment of inertia is a measure of a cross-section's resistance to torsional deformation and depends on the geometry of the cross-section. For a solid circular shaft, J can be calculated as $\frac{\pi d^4}{32}$, where d is the diameter of the shaft.

Both axial and torsional deformations are critical in the design and analysis of structural components. Engineers must ensure that the materials and structures they design can withstand the expected loads without exceeding allowable deformation limits, which could compromise structural integrity or functionality. For instance, excessive axial elongation in a bridge cable could lead to sagging, while excessive torsional deformation in a drive shaft could result in misalignment and mechanical failure.

The interplay between axial and torsional loads is also a consideration in composite materials and structures where different materials are combined. The differing material properties can lead to complex stress and strain distributions under combined loading conditions. Advanced computational methods, such as finite element analysis (FEA), are often employed to accurately predict the behavior of such materials and structures under various loading scenarios.

Bending and Thermal Deformations

Deflection due to bending is a critical aspect of structural analysis, particularly in the context of beams and other structural elements subjected to transverse loads. The bending moment induced by such loads causes the beam to deform, a phenomenon that can be quantified by examining the curvature of the beam along its length. According to the Euler-Bernoulli beam theory, the relationship between the bending moment M and the curvature κ of a beam is given by $M = EI\kappa$, where E is the modulus of elasticity of the material and I is the moment of inertia of the beam's cross-section about the neutral axis. The curvature, defined as the reciprocal of the radius of curvature R, $\kappa = 1/R$, is a measure of how much a beam deflects under a load. The deflection y at any point along the beam can be determined by integrating the curvature twice with respect to the length x, considering the boundary conditions of the beam.

Thermal effects on deformation must also be considered, as temperature changes can cause materials to expand or contract. This thermal expansion or contraction can add to the stresses and strains experienced by a material, influencing its overall deformation under load. The linear thermal expansion coefficient α characterizes how the size of an object changes with a change in temperature. For a beam experiencing a uniform temperature change ΔT, the thermal strain $\epsilon_{thermal}$ can be expressed as $\epsilon_{thermal} = \alpha \Delta T$. This strain, when multiplied by the modulus of elasticity E, gives the thermal stress induced in the beam. In the context of deformation analysis, the effect of thermal expansion or contraction must be superimposed on the mechanical deformation due to loads. This superposition principle allows for the combined effects of mechanical and thermal loads to be analyzed together, providing a more comprehensive understanding of the beam's behavior.

The principle of superposition is particularly useful in complex loading scenarios where multiple loads and effects interact. It states that the total deformation of a structure under several actions is equal to the sum of the deformations caused by each action taken individually, provided the system remains linear and the deformations are small. This principle simplifies the analysis of structures subjected to various types of loads, including bending moments, shear forces, axial loads, and thermal effects. By breaking down the complex problem into simpler parts, each can be solved independently using the appropriate formulas and methods. The results are then summed to obtain the overall deformation, allowing engineers to predict the behavior of structures under real-world conditions accurately.

In applying these concepts to the FE Civil Exam preparation, it is essential to grasp the underlying principles of bending and thermal deformations and to be proficient in applying the superposition principle to solve complex problems. Understanding how to calculate the moment of inertia for various cross-sections, how to integrate to find deflection curves, and how to account for thermal effects in structural elements are foundational skills that will aid in the analysis and design of safe, efficient, and durable structures in civil engineering projects.

Combined Stresses and Mohr's Circle

Combined Stresses and Transformations

In the realm of mechanics of materials, understanding the concept of combined stresses is crucial for analyzing the behavior of materials under various loading conditions. This analysis is foundational for ensuring the structural integrity and safety of engineering designs. The principle of superposition of stresses, stress transformations, and the methodology for finding principal stresses are key components of this analysis.

The principle of superposition asserts that the total stress at a point in a material under multiple loads is the algebraic sum of the stresses caused by each load independently. This principle is applicable under linear elastic conditions, where the material's response to stress is directly proportional to the applied load. For instance, if a beam is subjected to both axial stress due to a tensile load and shear stress due to a transverse load, the total stress experienced by the beam at any point can be calculated by summing the individual stresses from each load.

Stress transformation equations, often represented in matrix form, are used to calculate stresses on planes at any orientation from the known stresses on perpendicular planes. This is particularly useful when the orientation of the material or component under analysis does not align with the principal axes of loading. The transformation equations for plane stress conditions are derived from the equilibrium of a differential element and involve the normal and shear stresses on the x and y axes, σ_x, σ_y, and τ_{xy}, as well as the angle of rotation, θ, from the original coordinate system. The transformed stresses, $\sigma_{x'}$, $\sigma_{y'}$, and $\tau_{x'y'}$, can be calculated using:

$$\sigma_{x'} = \frac{\sigma_x + \sigma_y}{2} + \frac{\sigma_x - \sigma_y}{2}\cos(2\theta) + \tau_{xy}\sin(2\theta)$$

$$\sigma_{y'} = \frac{\sigma_x + \sigma_y}{2} - \frac{\sigma_x - \sigma_y}{2}\cos(2\theta) - \tau_{xy}\sin(2\theta)$$

$$\tau_{x'y'} = -\frac{\sigma_x - \sigma_y}{2}\sin(2\theta) + \tau_{xy}\cos(2\theta)$$

These equations facilitate the analysis of stress components on any inclined plane within the stressed body, enabling engineers to understand the stress distribution more comprehensively.

Finding principal stresses is a critical step in assessing the maximum and minimum stresses that a material experiences, which are pivotal in failure analysis. Principal stresses, denoted as σ_1 and σ_2, are the normal stresses on planes where shear stress is zero. These stresses correspond to the major and minor axes of the stress ellipse in Mohr's circle representation, a graphical method used to determine principal stresses and maximum shear stresses from the known state of stress on an element. The principal stresses can be calculated using:

$$\sigma_1, \sigma_2 = \frac{\sigma_x + \sigma_y}{2} \pm \sqrt{\left(\frac{\sigma_x - \sigma_y}{2}\right)^2 + \tau_{xy}^2}$$

This equation highlights that the principal stresses are independent of the shear stress magnitude and depend solely on the difference between the normal stresses and the orientation of the element. Identifying these stresses is essential for applying failure theories, such as the maximum normal stress theory or the von Mises stress criterion, which predict the onset of yielding or failure in materials under complex loading conditions.

The analysis of combined stresses through the principles of superposition, stress transformation, and the determination of principal stresses is fundamental in the field of mechanics of materials. This analysis enables engineers to predict the behavior of materials and structures under various loading scenarios, ensuring their performance and safety in real-world applications.

Mohr's Circle: Construction, Interpretation, Applications

Mohr's Circle is a graphical representation used to determine and visualize the state of stress at a point in a material under combined loading conditions. It simplifies the process of finding principal stresses, maximum shear stresses, and the orientation of the stress elements, which are critical for assessing the material's failure criteria under complex loading scenarios. The construction of Mohr's Circle and the interpretation of its results are pivotal for engineers to predict the behavior of materials and ensure the structural integrity of engineering designs.

To construct Mohr's Circle, one must first identify the normal (σ) and shear (τ) stresses acting on the x and y axes of the element in question. The circle is then drawn with a center at $\frac{\sigma_x + \sigma_y}{2}$ on the horizontal axis, which represents the average normal stress, and the origin of the vertical axis representing the shear stress. The radius of the circle, R, is determined by the equation $R = \sqrt{\left(\frac{\sigma_x - \sigma_y}{2}\right)^2 + \tau_{xy}^2}$, which encapsulates the relationship between the differential normal stress and the shear stress. This radius effectively represents the maximum shear stress that the element can experience.

The points where the circle intersects the horizontal axis correspond to the principal stresses, σ_1 and σ_2, which are the maximum and minimum normal stresses the element undergoes, with no shear stress present. These principal stresses are crucial for evaluating the material's susceptibility to failure under normal stress criteria, such as the maximum normal stress theory or the Coulomb-Mohr criterion. The angle θ_p, which represents the orientation of the principal stresses with respect to the original coordinate system, can be found by drawing a line from the center of the circle to the point of interest on the circle's circumference. The angle between this line and the horizontal axis, when doubled, gives the physical orientation of the principal planes.

Interpreting the results obtained from Mohr's Circle requires understanding that the circle itself represents all possible states of stress for a given point under the applied loading conditions. By analyzing the circle, engineers can determine not only the magnitudes of the principal and shear stresses but also the directions in which these stresses act. This analysis is instrumental in designing components that can withstand the expected loading without failure, ensuring the safety and reliability of civil engineering structures.

Applications of Mohr's Circle extend beyond mere failure prediction. It is also used in the analysis of stress transformations, allowing engineers to assess the stress state on inclined planes within a material. This capability is particularly useful in geotechnical engineering for evaluating soil stability and in mechanical engineering for analyzing the stress state in rotating machinery components. Furthermore, Mohr's Circle provides insights into the ductile or brittle failure of materials by comparing the principal stresses to the material's yield criteria.

In the context of the FE Civil Exam, understanding how to construct and interpret Mohr's Circle equips candidates with the ability to solve complex problems related to stress analysis and material failure. Mastery of this topic not only aids in passing the exam but also lays a solid foundation for future professional practice, where the principles of mechanics of materials are applied to ensure the structural integrity and longevity of engineering projects.

Chapter 7: Materials

Mix Design of Concrete and Asphalt

Concrete Mix Design Essentials

The design of concrete mix plays a pivotal role in ensuring the structural integrity, durability, and performance of concrete structures. Central to this process is the understanding and application of the water-cement ratio, admixtures, and aggregate proportions, which collectively determine the concrete's strength and workability.

The water-cement ratio (W/C) is a critical factor in concrete mix design, fundamentally influencing the hydration process, and thus, the final strength and durability of the concrete. The W/C ratio is defined as the weight of water divided by the weight of cement in a concrete mix. A lower W/C ratio leads to higher strength and durability but may reduce workability. The American Concrete Institute (ACI) provides guidelines for selecting an appropriate W/C ratio based on the required concrete strength, exposure conditions, and aggregate size. For instance, a W/C ratio of 0.45 is commonly recommended for concrete exposed to severe weathering conditions, aiming for a balance between strength, durability, and workability.

Admixtures are added to the concrete mix to modify its properties to suit specific construction needs. These can be broadly categorized into chemical admixtures and mineral admixtures. Chemical admixtures, such as superplasticizers, retarders, accelerators, and air-entraining agents, are used to enhance workability, control setting time, improve strength, and introduce air voids for freeze-thaw resistance, respectively. Mineral admixtures, including fly ash, silica fume, and slag cement, contribute to the concrete's strength and durability by participating in the hydration process and reducing the W/C ratio through their pozzolanic or hydraulic activities. The selection and proportioning of admixtures must be carefully optimized based on the desired concrete performance characteristics, environmental exposure conditions, and compatibility with available cements and aggregates.

Aggregate proportions, encompassing both coarse and fine aggregates, are determined based on their size, shape, texture, and grading. Aggregates occupy 60-75% of the concrete volume, making their properties critical in defining the mix's overall performance. The grading of aggregates influences the concrete's workability, pumpability, and durability. Well-graded aggregates, with a mix of different sizes, tend to produce a denser and more workable mix, reducing the voids that need to be filled with cement paste and thus optimizing the W/C ratio. The ACI provides guidelines for aggregate selection and grading to achieve the desired concrete strength and workability.

Achieving the desired concrete strength and workability necessitates a meticulous balance between the W/C ratio, admixtures, and aggregate proportions. This balance is guided by the principles of mix design, which include the Abrams' law for the W/C ratio's impact on strength, the use of admixtures for targeted property enhancements, and the optimization of aggregate grading for workability and compactness. Trial mixes are essential in validating the designed mix's performance, ensuring it meets the specified strength requirements and workability for efficient placement and compaction.

In practice, the concrete mix design process involves iterative adjustments and testing to fine-tune the mix proportions, ensuring compliance with design specifications and performance criteria. This process underscores the importance of a thorough understanding of material properties, environmental conditions, and structural requirements in crafting a concrete mix that meets the rigorous demands of modern construction projects.

Asphalt Mix Design Principles

Asphalt mix design is a meticulous process that involves the careful selection and combination of aggregates and binder to achieve a final product that meets specific performance criteria. The primary goal is to create a mix that is durable, resistant to deformation, and capable of withstanding various loads and environmental conditions. This section delves into the critical components of asphalt mix design, including **aggregate gradation**, **binder content**, **performance grading**, and strategies for enhancing **durability** and **load resistance**.

Aggregate Gradation plays a pivotal role in the mix design of asphalt. It refers to the distribution of particle sizes within the mix and is crucial for achieving the desired density, stability, and voids in the mineral aggregate (VMA). A well-graded aggregate blend, characterized by a wide range of sizes from coarse to fine, ensures a tighter packing and more efficient use of the binder. The Superpave system specifies aggregate gradation requirements that aim to optimize the skeleton of the asphalt mix, enhancing its resistance to permanent deformation and fatigue. The gradation is typically evaluated using the 0.45 power gradation chart, where the ideal curve represents the optimal distribution of particle sizes for maximum density and performance.

Binder Content is another critical factor in asphalt mix design. The binder, usually asphalt cement, serves as the glue that holds the aggregate particles together. The optimal binder content is essential for ensuring sufficient coating of aggregates, workability during construction, and long-term performance of the pavement. It is determined based on the balance between achieving adequate film thickness around aggregate particles and preventing excessive binder that leads to bleeding or rutting. The selection of binder content typically involves laboratory tests, such as the Marshall or Superpave mix design methods, to evaluate the mix's stability, flow, and voids in the total mix (VTM) at various binder contents.

Performance Grading of the binder is a system developed under the Strategic Highway Research Program (SHRP) to classify asphalt binders based on their performance in expected pavement temperatures. This system, known as the Performance Grade (PG) system, identifies binders by their high and low-temperature performance limits, e.g., PG 64-22, where 64 represents the high-temperature limit ($^\circ C$) and -22 the low-temperature limit. The PG system ensures that the selected binder is appropriate for the climatic conditions of the project location, contributing to the asphalt mix's durability and resistance to thermal cracking, rutting, and fatigue.

Designing for **Durability and Load Resistance** involves the integration of mix components that can withstand the stresses imposed by traffic loads and environmental conditions over the pavement's lifespan. Durability is enhanced by optimizing the binder film thickness to prevent oxidation and aging of the asphalt mix. Load resistance, particularly rutting resistance, is achieved by selecting a binder with suitable high-temperature performance and designing a mix with a stable aggregate structure. The use of modifiers, such as polymers and fibers, can further improve the mix's performance by increasing its stiffness and resistance to deformation.

The asphalt mix design process is a complex interplay of material selection and proportioning to meet specific engineering and performance criteria. By carefully controlling aggregate gradation, binder content, and performance grading, engineers can design asphalt mixes that offer superior durability, load resistance, and overall pavement performance. This approach ensures that the resulting pavement structure can support the demands of traffic and environmental conditions, providing a safe and smooth driving experience for the public.

Test Methods and Specifications

Concrete and Aggregates: Explain slump tests, compressive strength tests, gradation, and specific standards for aggregate properties.

The slump test is a crucial method used to determine the workability of fresh concrete by measuring its consistency. This test involves filling a conical mold with concrete and then removing the mold to observe how much the concrete slumps or settles under its own weight. The degree of slump is indicative of the concrete's workability; a higher slump value signifies a more workable mix, while a lower slump indicates a stiffer, less workable concrete.

In addition to the slump test, the compressive strength test is employed to evaluate the concrete's ability to withstand applied loads. This test measures the maximum load that a concrete specimen can bear before failure, providing essential data for structural design and safety assessments.

Furthermore, the gradation test is utilized to assess the particle size distribution of aggregates used in concrete. Proper gradation is vital for achieving optimal packing and minimizing voids within the concrete mix, which directly influences the strength and durability of the final product.

Together, these tests form a comprehensive approach to ensuring the quality and performance of concrete in construction applications.

Metals, Asphalt, and Wood: Discuss tensile tests for metals, asphalt binder tests (penetration/viscosity), and mechanical tests for wood.

The evaluation of materials such as metals, asphalt binders, and wood is crucial in engineering applications, particularly in construction and structural design.

The **tensile test** is a fundamental method used to determine the tensile strength of metals. This test involves applying a uniaxial force to a specimen until it fractures. The results provide valuable information about the material's ability to withstand tension, which is essential for ensuring structural integrity in engineering applications.

For asphalt binders, the **penetration or viscosity test** is employed to assess their viscosity. This test measures the resistance of the asphalt to flow under specific conditions, which is critical for understanding how the material will perform in various temperatures and loading conditions. The viscosity of asphalt affects its workability and durability, making this test vital for pavement design and maintenance.

In the case of wood, the **flexural test** is utilized to evaluate its mechanical properties, particularly its bending strength and stiffness. During this test, a load is applied to the wood until it bends or breaks, allowing engineers to determine how the material will behave under load. Understanding the flexural properties of wood is essential for its use in structural applications, ensuring that it can support the required loads without failure.

Together, these tests provide a comprehensive understanding of the mechanical properties of metals, asphalt, and wood, enabling engineers to make informed decisions in material selection and structural design.

Physical and Mechanical Properties

Basic Properties of Construction Materials

Understanding the basic properties of materials such as metals, concrete, aggregates, asphalt, and wood is crucial for civil engineers to design, construct, and maintain infrastructure effectively.

These properties include **density**, **porosity**, **elasticity**, **toughness**, and **thermal properties**, each playing a pivotal role in determining the suitability of a material for specific engineering applications.

Density (ρ) is a fundamental property that reflects the mass per unit volume of a material ($\rho = \dfrac{m}{V}$), typically expressed in kilograms per cubic meter (kg/m^3) or pounds per cubic foot (lb/ft^3). Metals, known for their compact atomic structure, exhibit higher densities, which influence their weight and strength characteristics. Concrete, a composite material, has a variable density depending on its composition, including the types of aggregates used. Aggregates themselves vary widely in density, affecting the overall density of concrete. Asphalt, used primarily in road construction, has a density that impacts its compaction and performance. Wood, with its cellular structure, has a lower density compared to other construction materials, which contributes to its excellent strength-to-weight ratio.

Porosity represents the volume fraction of void spaces within a material, affecting its ability to absorb water and other fluids. This property is critical for materials like concrete and aggregates, where porosity influences durability, strength, and resistance to freeze-thaw cycles. Lower porosity in metals and asphalt leads to higher strength and durability, making them ideal for structural and paving applications, respectively.

Elasticity is the ability of a material to return to its original shape after removing the force causing deformation. Described by the modulus of elasticity or Young's modulus (E), it quantifies the stiffness of a material. Metals typically have high values of E, indicating less deformation under load, essential for structural applications. Concrete and asphalt, while less elastic than metals, must still exhibit sufficient elasticity to withstand load without cracking. Wood's elasticity varies with the grain direction, a factor that engineers must consider in design and construction.

Toughness measures the ability of a material to absorb energy and plastically deform without fracturing. It is crucial for materials subjected to impact or shock loads. Metals often exhibit high toughness, making them suitable for dynamic loading conditions. Concrete, while strong under compression, has lower toughness, necessitating reinforcement with steel to improve its performance under tensile stresses. Asphalt's toughness is optimized for road surfaces to resist cracking under traffic loads. Wood's natural toughness varies with species and moisture content, influencing its use in construction.

Thermal properties, including thermal conductivity, specific heat capacity, and thermal expansion coefficient, determine how materials respond to temperature changes. Metals, with high thermal conductivity, quickly transfer heat, impacting their use in thermal management systems. Concrete and asphalt have moderate thermal properties that affect their performance in varying climates. Wood's low thermal conductivity makes it an excellent insulator, beneficial for energy-efficient building designs.

Strength Properties of Construction Materials

Understanding the strength properties of construction materials is fundamental for civil engineers to ensure the safety, reliability, and longevity of structures. These properties include compressive strength, tensile strength, shear strength, fatigue resistance, and durability, each of which plays a critical role in the material's performance under various loads and environmental conditions.

Compressive strength is the capacity of a material to withstand loads tending to reduce size. For concrete, the American Concrete Institute (ACI) specifies the standard test method for compressive strength, which involves crushing cylindrical concrete specimens in a compression-testing machine. The compressive strength of concrete is a primary consideration in the design and construction of buildings and infrastructure. Metals, such as steel, also exhibit high compressive strengths, making them ideal for use in high-load-bearing applications.

Tensile strength, on the other hand, measures the force required to pull something such as rope, wire, or a structural beam to the point where it breaks. The tensile strength of steel is a critical parameter in the design of structures, as it ensures that elements can withstand the forces exerted upon them without failure. The testing of tensile strength is conducted by applying a tensile force to a sample until it fractures, and the maximum applied force is recorded. The ASTM A370 is a commonly referenced standard for the tensile testing of metals.

Shear strength is the ability of a material to resist forces that can cause the internal structure of the material to slide against itself. This property is crucial in the design of fasteners, beams, and foundations that are subjected to lateral loads. The shear strength of a material is typically tested by applying a force parallel to the surface of the material. In soils, for example, the shear strength is a key factor in the design of retaining walls and slopes, ensuring stability against sliding failures.

Fatigue resistance refers to a material's ability to withstand repeated loading and unloading cycles without failing. Materials that are subjected to cyclic stresses, such as bridge components and aircraft structures, must exhibit high fatigue resistance to avoid premature failure. The fatigue life of a material is determined through laboratory testing, where samples are subjected to controlled stress cycles until failure occurs. The S-N curve, or Wöhler curve, is a fundamental graph in materials science that plots the amplitude of cyclic stress (S) against the logarithm of the number of cycles to failure (N).

Durability, in the context of construction materials, refers to the ability of a material to withstand wear, pressure, or damage over time, maintaining its integrity and functionality. Factors affecting the durability of materials include environmental conditions, such as exposure to moisture, chemicals, and temperature fluctuations, as well as the material's inherent properties. For instance, the durability of concrete can be enhanced through proper mix design, curing, and the use of admixtures to resist freeze-thaw cycles and chemical attacks.

The selection of materials for any civil engineering project must consider these strength properties to ensure that the constructed facility can withstand the intended loads and environmental conditions over its expected lifespan. By understanding and applying knowledge of compressive, tensile, shear strengths, fatigue resistance, and durability, engineers can design structures that not only meet safety standards but also offer longevity and reliability, contributing to sustainable infrastructure development.

Chapter 8: Fluid Mechanics

Flow Measurement

Measurement Methods: Venturi, Orifice, Pitot Applications

Venturi meters, orifice plates, and Pitot tubes are fundamental devices used in the field of fluid mechanics for the measurement of flow rate and velocity, each employing distinct principles of operation to achieve accurate readings under various conditions. Understanding the mechanics, applications, and limitations of these instruments is crucial for engineers to effectively design, analyze, and optimize fluid systems.

Venturi meters operate on the principle of the Venturi effect, which states that the velocity of a fluid increases as the cross-sectional area of the flow path decreases, resulting in a corresponding decrease in fluid pressure. This phenomenon is described mathematically by the Bernoulli equation, which relates the pressure, velocity, and elevation head at two points along the flow path. The Venturi meter consists of a converging section that accelerates the fluid, a throat where the cross-sectional area is minimum and velocity is maximum, and a diverging section where the fluid decelerates. The differential pressure between the upstream side and the throat is measured and used to calculate the flow rate using the continuity and Bernoulli equations. Venturi meters are favored for their high accuracy, low head loss, and suitability for measuring the flow rates of liquids, gases, and steam in large pipes.

Orifice plates are simpler and more cost-effective devices compared to Venturi meters, consisting of a flat plate with a central hole (orifice) inserted into the flow stream. As fluid passes through the orifice, its velocity increases, creating a pressure differential between the upstream and downstream sides of the plate. The flow rate can then be determined from this pressure drop, the orifice area, and the fluid properties, applying the Bernoulli equation and flow coefficients that account for energy losses and non-ideal conditions. Orifice plates are versatile and widely used but introduce significant head loss and require regular calibration to maintain accuracy, especially in dirty or viscous fluids.

Pitot tubes, named after the French engineer Henri Pitot, measure fluid velocity at a point within a flow stream. A Pitot tube consists of a tube pointing directly into the fluid flow, with an opening at the end that allows the fluid to enter. This tube is connected to a manometer or pressure sensor that measures the stagnation pressure (the pressure when the fluid is brought to a complete stop). The static pressure of the fluid is measured separately, and the difference between the stagnation and static pressures, known as the dynamic pressure, is used to calculate the fluid velocity based on Bernoulli's principle. Pitot tubes are particularly useful for measuring airspeed in aviation and flow velocity in ducts and pipes. They are simple, inexpensive, and can be inserted directly into existing systems with minimal disruption to the flow.

Each of these measurement methods has its own set of advantages, limitations, and applications, making them indispensable tools in the arsenal of civil engineers. The choice of device depends on factors such as the fluid type, flow conditions, accuracy requirements, and economic considerations. Proper selection, installation, and maintenance of Venturi meters, orifice plates, and Pitot tubes are essential for obtaining reliable measurements and ensuring the efficient operation of fluid systems in various engineering applications.

Practical Measurement Techniques

Calibration, error analysis, and the selection of appropriate measurement techniques are critical components in the accurate assessment of flow rates within fluid mechanics. These processes ensure the reliability and accuracy of data collected using devices such as Venturi meters, orifice plates, and Pitot tubes, which are fundamental in designing and analyzing fluid systems.

Calibration involves adjusting the measurement device to conform to a known standard or accuracy. For Venturi meters and orifice plates, calibration may include verifying the differential pressure measurement system against a known pressure source. Pitot tubes require calibration against a standard velocity field, often in a calibration tunnel, to ensure accurate velocity measurements. The calibration process accounts for deviations in manufacturing, installation variances, and changes in the measurement environment that may affect the device's accuracy. Regular calibration schedules are essential, as they compensate for wear and tear and environmental impacts over time.

Error analysis in flow measurement encompasses the identification and quantification of potential sources of error. These errors can be systematic, arising from inaccuracies in the measurement system itself, or random, resulting from external variables such as fluctuations in fluid properties or environmental conditions. For instance, the placement of an orifice plate or a Pitot tube in a flow stream where turbulent eddies or swirls are present can introduce significant measurement errors. Error analysis involves statistical methods to estimate the uncertainty in measurements, guiding the engineer in understanding the reliability of the data collected.

Selecting the **appropriate measurement technique** for a specific system requires a thorough understanding of the fluid's properties, the flow conditions, and the system's geometric constraints. For example, Venturi meters are preferred in applications where low-pressure drop and high accuracy are critical, and the fluid is clean and non-viscous. Orifice plates, being simpler and less expensive, are suitable for a wide range of applications but may not be the best choice for measuring flows with high particulate content or highly viscous fluids due to the significant energy loss and potential for clogging. Pitot tubes offer the advantage of point velocity measurement and are invaluable in applications requiring velocity profiling across a flow section. However, their accuracy is highly dependent on the correct alignment with the flow direction and the flow's stability.

In practical applications, engineers must also consider the **installation effects** that can influence measurement accuracy. Proper alignment with the flow direction, sufficient straight pipe lengths upstream and downstream of the measurement device, and avoidance of flow disturbances are critical factors. The presence of bends, valves, or fittings near the measurement site can introduce flow distortions that significantly affect the accuracy of the measurement.

Ultimately, the selection of flow measurement devices and techniques must be guided by a comprehensive analysis of the system requirements, including accuracy, cost, maintenance, and the specific characteristics of the fluid and flow. By meticulously calibrating devices, rigorously analyzing potential errors, and judiciously selecting the most suitable measurement technique, engineers can ensure the collection of reliable data, which is paramount for the effective design, optimization, and operation of fluid systems.

Fluid Properties

Basic Fluid Properties

Density, viscosity, surface tension, and compressibility are fundamental fluid properties that significantly influence the behavior of fluids in various engineering contexts. Understanding these properties is crucial for civil engineers to design and analyze systems involving fluid flow, such as water distribution networks, wastewater treatment facilities, and flood control systems.

Density (ρ) is defined as the mass per unit volume of a fluid and is a primary characteristic that affects fluid statics and dynamics. It is expressed in kilograms per cubic meter (kg/m^3) or pounds per cubic foot (lb/ft^3). The density of a fluid influences buoyancy, pressure, and energy calculations. For instance, the hydrostatic pressure at a point within a fluid is directly proportional to the fluid's density and the depth of the point below the surface, as described by the equation $P = \rho g h$, where P is the pressure, g is the acceleration due to gravity, and h is the depth.

Viscosity (μ) is a measure of a fluid's resistance to flow and deformation by shear or tensile stress. It is a critical factor in the analysis of fluid flow, affecting the velocity profile and the energy loss in piping systems. Viscosity is quantified in pascal-seconds ($Pa \cdot s$) or poise (P), with water at $20°C$ having a dynamic viscosity of approximately $1.002 \times 10^{-3} Pa \cdot s$. The Navier-Stokes equations, which describe the motion of viscous fluid substances, incorporate viscosity to predict the flow velocity, pressure, and shear stresses in a fluid.

Surface tension (σ) is the elastic tendency of fluid surfaces to acquire the least surface area possible. This property is significant at the interface between two fluids, such as water and air, and affects phenomena like capillarity and droplet formation. Surface tension is measured in newtons per meter (N/m) and plays a vital role in the design of structures that interact with fluids, influencing the wetting and spreading of liquids on solid surfaces. The Young-Laplace equation, which relates the pressure difference across a fluid interface to the surface tension and curvature of the interface, demonstrates the impact of surface tension on fluid mechanics.

Compressibility (β) describes the change in volume of a fluid in response to a change in pressure, highlighting the fluid's ability to be compressed. It is inversely related to the bulk modulus of elasticity (K), with $\beta = 1/K$, and is particularly relevant for gases, which are highly compressible compared to liquids. Compressibility affects the speed of sound in a fluid, the density changes in fluid flow, and the shock waves formation. For engineering applications involving high-speed gas flow, such as air ventilation systems and aerodynamics, understanding compressibility is essential.

These basic fluid properties—density, viscosity, surface tension, and compressibility—form the foundation for analyzing and solving a wide range of civil engineering problems involving fluid mechanics. By applying principles that incorporate these properties, engineers can design more efficient, effective, and sustainable systems for managing water resources, controlling floods, treating wastewater, and constructing hydraulic structures.

Specialized Fluid Properties in Engineering

In the realm of fluid mechanics, understanding the specialized properties of fluids, such as the distinction between Newtonian and non-Newtonian fluids, vapor pressure, and the phenomenon of cavitation, is crucial for engineers to design and analyze systems effectively. These properties significantly influence the behavior of fluids under various conditions and, therefore, have direct implications on engineering solutions.

Newtonian fluids are characterized by a constant viscosity that does not change with the rate of deformation or shear rate. This means that the fluid's resistance to flow remains constant regardless of the force applied to it. Water, air, and most gases are classic examples of Newtonian fluids. The shear stress τ in a Newtonian fluid is directly proportional to the shear rate $\dot{\gamma}$, represented by the equation $\tau = \mu \dot{\gamma}$, where μ is the dynamic viscosity of the fluid. This linear relationship is pivotal in simplifying the analysis and design of systems involving these fluids.

Conversely, non-Newtonian fluids exhibit a viscosity that changes with the rate of deformation. This category includes fluids whose flow behavior cannot be described by a single constant viscosity. Examples of non-Newtonian fluids include slurries, suspensions, polymers, blood, and many other biological fluids. The shear stress-shear rate relationship for non-Newtonian fluids is not linear, and these fluids are further classified into subcategories such as pseudoplastic, dilatant, Bingham plastic, and thixotropic fluids, each demonstrating unique flow characteristics. Engineers must understand the specific behavior of non-Newtonian fluids to predict their behavior accurately in processes and equipment design.

Vapor pressure is another critical fluid property, representing the pressure at which a liquid and its vapor are in equilibrium at a given temperature. It is a fundamental concept in understanding boiling, evaporation, and condensation processes. The vapor pressure of a fluid increases with temperature, and when the vapor pressure equals the atmospheric pressure, the liquid boils. This property is essential in the design of heat exchangers, boilers, and refrigeration systems, where phase change processes are integral.

Cavitation is a phenomenon that occurs when the local pressure in a fluid falls below the vapor pressure, leading to the formation of vapor bubbles within the fluid. These bubbles can collapse violently when they move to regions of higher pressure, causing shock waves that can damage machinery, such as pumps, turbines, and propellers. Understanding cavitation is vital for engineers to design equipment that minimizes its occurrence and adverse effects. The prediction and control of cavitation require a thorough understanding of the fluid's vapor pressure and the system's pressure dynamics.

The specialized properties of fluids, including the behavior of Newtonian and non-Newtonian fluids, vapor pressure, and cavitation, are fundamental concepts in fluid mechanics that have significant implications in engineering practice. Mastery of these concepts enables engineers to design more efficient, reliable, and safe systems across a wide range of applications

Fluid Statics

Pressure in Static Fluids

In the realm of fluid mechanics, **hydrostatic pressure** is a fundamental concept that describes the pressure exerted by a fluid at rest due to the force of gravity. It is calculated as $P = \rho g h$, where P represents the pressure, ρ is the fluid density, g is the acceleration due to gravity, and h is the height of the fluid column above the point of measurement. This equation underscores the direct relationship between the depth of a fluid and the pressure exerted by that fluid, illustrating how pressure increases with depth within a fluid body due to the weight of the fluid above.

Pressure head, another critical concept in fluid statics, refers to the height of a fluid column that results in a specific pressure at the base of the column. It is expressed as $h = \dfrac{P}{\rho g}$, effectively describing the equivalent height of a fluid column that would exert the same pressure as the actual pressure at a given point. This concept is pivotal in hydraulic engineering, enabling engineers to conceptualize and calculate the pressure in terms of fluid column height, facilitating the design and analysis of water supply systems, dams, and other hydraulic structures.

Pascal's principle posits that pressure applied to an enclosed fluid is transmitted undiminished to every part of the fluid, as well as to the walls of its container. This principle can be mathematically represented as $P_1 = P_2$, indicating that the pressure at one point in a closed system equals the pressure at another point, provided there is no difference in elevation between the two points. Pascal's principle forms the basis for hydraulic systems, such as brakes and lifts, where a small force applied at one point generates a much larger force at another point, enabling the efficient transmission of power within the system.

Understanding these concepts is crucial for engineers to design systems that can withstand the forces exerted by static fluids. For instance, calculating the hydrostatic pressure at various depths is essential for determining the structural requirements of underwater tunnels, submarine hulls, and the foundations of dams. Similarly, knowledge of pressure head is indispensable for the design of water distribution networks, ensuring that water can be efficiently delivered to elevated locations. Pascal's principle, with its implications for force multiplication, is fundamental in the design of machines and devices that rely on hydraulic power for operation.

The application of these principles extends beyond theoretical calculations, impacting the practical aspects of engineering projects. For example, in the design of a dam, engineers must calculate the hydrostatic pressure at various depths to ensure that the structure can withstand the immense forces exerted by the water. In water treatment facilities, understanding pressure head allows engineers to design systems that ensure the adequate flow and distribution of water through the treatment process. In the realm of machinery, Pascal's principle is applied in the design of hydraulic presses, which are used in various manufacturing processes to mold, form, or compress materials.

In summary, hydrostatic pressure, pressure head, and Pascal's principle are foundational concepts in fluid mechanics that enable engineers to analyze and design a wide range of systems and structures that interact with static fluids. Mastery of these concepts is essential for the effective application of fluid mechanics principles in engineering solutions, ensuring the safety, reliability, and efficiency of hydraulic systems and structures.

Applications of Statics

Buoyancy, a fundamental principle in fluid statics, is governed by Archimedes' principle, which states that any body fully or partially submerged in a fluid at rest is acted upon by an upward, or buoyant, force. The magnitude of this force is equal to the weight of the fluid displaced by the body. Mathematically, the buoyant force (F_b) can be expressed as $F_b = \rho_f \cdot g \cdot V$, where ρ_f is the density of the fluid, g is the acceleration due to gravity, and V is the volume of fluid displaced by the submerged part of the body. This principle is crucial for the design and analysis of ships, submarines, and floating structures, ensuring they remain afloat and stable under various conditions.

Stability of floating bodies is another critical application of statics in fluid mechanics, closely related to buoyancy. A floating body is stable if, when it is displaced, it returns to its original position. The stability is determined by the relationship between the center of gravity (CG) and the center of buoyancy (CB). The center of gravity is the point where the weight of the body acts, and the center of buoyancy is the point where the force of buoyancy acts. For a body to be stable, the center of buoyancy must be directly above the center of gravity when the body is slightly tilted. If the body is tilted and the center of buoyancy moves to the side of the center of gravity, a righting moment is created that tends to return the body to its original position, indicating stable equilibrium. Engineers use these concepts to design vessels and structures that not only float but maintain their orientation and stability in water.

Manometers are devices used to measure the pressure of a fluid by balancing the fluid column against a known pressure. They are an essential tool in fluid statics for measuring pressure differences in fluids at rest. A simple manometer consists of a U-shaped tube filled with a fluid of known density, with one end connected to the point of interest in the fluid system and the other end open to the atmosphere or another reference pressure. The pressure at the point of interest is determined by the height difference (Δh) of the fluid columns in the two arms of the U-tube, given by the equation $P = \rho \cdot g \cdot \Delta h$, where P is the pressure difference, ρ is the density of the manometric fluid, g is the acceleration due to gravity, and Δh is the height difference. Manometers are widely used in engineering applications to calibrate instruments, verify process pressures, and in laboratory experiments where accurate pressure measurements are required.

Essential skills for engineers in fluid mechanics include the principles of buoyancy, the stability of floating bodies, and the operation of manometers. These principles provide a theoretical basis for analyzing and designing systems that interact with fluids, while also ensuring that the practical implementation of these systems is feasible, safe, and efficient. Proficiency in these concepts enables engineers to address complex challenges in the design of marine vessels, hydraulic structures, and various fluid systems, thereby ensuring their functionality and safety in real-world applications.

Energy, Impulse, and Momentum of Fluids

Energy Principles in Fluid Systems

Bernoulli's equation is a cornerstone of fluid mechanics, encapsulating the principle of energy conservation in fluid flow. It asserts that for an incompressible, frictionless fluid, the total mechanical energy along a streamline remains constant. Mathematically, Bernoulli's equation is expressed as:

$$P + \frac{1}{2}\rho v^2 + \rho g h = \text{constant}$$

where P represents the pressure energy per unit volume, $\frac{1}{2}\rho v^2$ the kinetic energy per unit volume, and $\rho g h$ the potential energy per unit volume. Here, ρ denotes the fluid density, v the fluid velocity, g the acceleration due to gravity, and h the height above a reference point. This equation is instrumental in analyzing fluid flow problems where energy conversion between its forms occurs, such as in the design of venturi meters, pumps, and turbines.

Energy conservation in fluid systems is a fundamental concept, emphasizing that energy cannot be created or destroyed in an isolated system. The energy in a fluid system may transform from one form to another—potential energy to kinetic energy or vice versa—but the total energy remains constant. This principle is crucial in hydraulic engineering, enabling the prediction and analysis of fluid behavior in various engineering applications, including water supply, irrigation systems, and energy generation facilities.

Head loss is a critical aspect of real-world fluid flow, accounting for the reduction in mechanical energy as fluid moves through a system. This loss is primarily due to friction and can be categorized into major and minor losses. Major head loss occurs due to friction within the length of pipes and is calculated using the Darcy-Weisbach equation:

$$h_f = f \frac{L}{D} \frac{v^2}{2g}$$

where h_f is the friction head loss, f the friction factor, L the length of the pipe, D its diameter, v the velocity of the fluid, and g the acceleration due to gravity. Minor head losses, on the other hand, result from pipe fittings, bends, valves, and other components disrupting the flow, and are quantified using empirical formulas and coefficients specific to each type of fitting.

Understanding the implications of Bernoulli's equation, energy conservation, and head loss is essential for engineers to design efficient fluid systems. It enables the calculation of pressure drops, flow rates, and energy requirements, ensuring that systems operate within their intended parameters. For instance, in water distribution networks, accurately calculating head loss is vital for pump selection and determining the necessary pipe diameters to ensure adequate water pressure at the end-users. Similarly, in wastewater treatment plants, understanding these principles helps in designing processes that minimize energy consumption while maximizing treatment efficiency.

The application of these energy principles extends beyond traditional civil engineering projects. In environmental engineering, they are used to model natural water systems, predict pollutant dispersion, and design sustainable stormwater management solutions. In the context of renewable energy, engineers apply these principles to optimize the efficiency of hydroelectric power plants and develop innovative technologies such as tidal and wave energy converters.

Mastering Bernoulli's equation, the concept of energy conservation, and the mechanisms of head loss equips engineers with the tools to tackle complex fluid mechanics challenges. These principles form the foundation for designing systems that are not only functional and reliable but also sustainable and energy-efficient, aligning with the broader goals of environmental stewardship and resource conservation.

Impulse and Momentum in Fluid Dynamics

The principles of **impulse** and **momentum** play a pivotal role in the analysis of fluid dynamics problems, particularly when examining the forces exerted by fluids on bends and nozzles. The momentum equation, derived from Newton's second law, is fundamental in understanding how fluid motion relates to the forces acting on and within fluid systems.

The **momentum equation** for a fluid flowing through a control volume can be expressed as:

$$\sum \mathbf{F} = \frac{d}{dt}\int_{CV} \rho \mathbf{v}\, dV + \int_{CS} \rho \mathbf{v}(\mathbf{v} \cdot d\mathbf{A})$$

where $\sum \mathbf{F}$ represents the sum of external forces acting on the control volume, ρ is the fluid density, $\mathbf{v}$ is the fluid velocity vector, dV is the differential volume element, and $d\mathbf{A}$ is the differential area vector pointing outward over the control surface (CS). The first term on the right-hand side represents the rate of change of linear momentum within the control volume, while the second term accounts for the net outflow of linear momentum through the control surface.

In the context of **forces on bends and nozzles**, the momentum equation helps in calculating the resultant force required to change the direction or magnitude of the fluid velocity. For instance, in a pipe bend, the fluid changes direction, which according to the momentum principle, results in a force on the pipe bend. This force can be calculated by considering the momentum change of the fluid as it enters and exits the bend.

For a nozzle, the equation assists in determining the force exerted by the fluid as it accelerates or decelerates through the nozzle. The change in momentum of the fluid, attributed to the change in velocity as the fluid passes through the nozzle, directly relates to the force exerted on the nozzle.

Applications in fluid dynamics problems often involve analyzing the effects of these forces on system components. For example, in designing a water distribution network, understanding the forces at bends and junctions is crucial for ensuring the structural integrity of the pipes. Similarly, in pump and turbine design, the forces on nozzles influence the mechanical stress on the equipment and thus its durability and performance.

The **impulse-momentum principle** also finds application in solving problems related to jet propulsion, where the force produced by a jet issuing from a nozzle is a direct consequence of the momentum change of the fluid. This principle is essential in calculating the thrust generated by jet engines and waterjets.

In summary, the impulse and momentum equations provide a robust framework for analyzing and solving a wide array of fluid dynamics problems. By applying these principles, engineers can predict the forces exerted by fluids in motion, facilitating the design and analysis of systems such as pipelines, nozzles, and propulsion systems. Understanding and accurately applying these equations are crucial for ensuring the safety, efficiency, and reliability of fluid systems in various engineering applications.

Chapter 9: Surveying

Angles, Distances, and Trigonometry

Measuring Techniques in Surveying

In the realm of surveying, precise measurement of angles and distances forms the backbone of accurate land assessment and mapping. The methodologies employed for these measurements leverage advanced instruments and mathematical principles to ensure the highest degree of accuracy. Among these instruments, theodolites and total stations are paramount in their application, each serving a unique purpose in the surveying process.

Theodolites, intricate optical instruments, are utilized for the measurement of horizontal and vertical angles. Their precision is critical in delineating the boundaries of a parcel of land or in the construction of large infrastructure projects where angular accuracy is paramount. The operation of a theodolite involves aligning its sights on a target, with measurements read directly from its calibrated circles. These readings enable the calculation of angles which, when combined with distances measured, can accurately describe the layout of the land.

Total stations elevate the capabilities of traditional theodolites by integrating electronic distance measurement (EDM) technology, allowing for the simultaneous measurement of angles and distances. This integration not only streamlines the surveying process but also enhances accuracy. A total station emits a laser beam towards a prism held at the point to be measured. The time taken for the laser to return is then calculated and converted into a distance, factoring in variables such as signal velocity and atmospheric conditions. The comprehensive data captured by total stations, including angle, distance, and elevation information, is pivotal for creating detailed topographical maps and for the precise placement of structures.

The application of basic trigonometry in surveying further complements these measurements. Trigonometric principles enable the calculation of distances and angles from the data obtained through theodolites and total stations. For instance, the sine and cosine rules can be applied to calculate unknown distances or angles in a triangle when two sides and one angle, or two angles and one side, are known. This is particularly useful in triangulation, a method employed to determine the positions of points spread over large areas by dividing the region into a network of triangles.

Moreover, the concept of trigonometric leveling, where differences in elevation are determined using trigonometric functions based on angle measurements and horizontal distances, showcases the indispensable role of trigonometry in surveying. This method is especially beneficial in areas where traditional leveling techniques are impractical due to terrain or other constraints.

The synergy between sophisticated measuring instruments like theodolites and total stations, and the foundational principles of trigonometry, underscores the technical sophistication inherent in modern surveying practices. This combination not only ensures the precision required for contemporary engineering projects but also exemplifies the application of mathematical concepts in solving real-world challenges. Through the meticulous measurement of angles and distances, surveyors are able to translate the complexities of the physical world into accurate, usable data, underscoring the critical role of surveying in the planning and execution of engineering endeavors.

Applications of Triangulation and Error Adjustment

Triangulation and traversing are fundamental techniques in surveying that enable the accurate measurement of distances and the establishment of precise locations of points on the earth's surface. These methods, when applied correctly, can significantly enhance the accuracy of field measurements, which is crucial for a wide range of engineering projects.

Triangulation is a surveying method that determines the location of a point by forming triangles to it from known points. The primary principle behind triangulation is the mathematical concept that if one side and two angles of a triangle are known, the other two sides and angle can be calculated using trigonometric functions. In practical applications, triangulation begins with the establishment of a baseline of known length. From the endpoints of this baseline, angles to the unknown point are measured using theodolites or total stations. These measurements, along with the known baseline length, allow for the calculation of the distances from the known points to the unknown point, effectively determining its location. This method is particularly useful over long distances where direct measurement is impractical or impossible.

Traversing, on the other hand, involves the sequential measurement of distances and angles to establish a network of interconnected points. In a closed traverse, the path of measurement returns to the starting point, forming a polygon, which allows for the checking of measurements for consistency and accuracy. An open traverse does not return to the starting point and is used for extending control networks. Each leg of the traverse forms a side of a polygon, and the angles between each leg are measured. The lengths of the legs are determined using electronic distance measurement (EDM) equipment. The coordinates of each point in the traverse can then be calculated from the initial known point using the measured angles and distances.

Error adjustment is a critical component in both triangulation and traversing, as it ensures the accuracy and reliability of the survey data. Errors in surveying measurements can arise from a variety of sources, including instrument calibration, environmental conditions, and human factors. Systematic errors, which are predictable and consistent, can often be mitigated through calibration and correction procedures. Random errors, which are unpredictable and vary in magnitude and direction, are addressed through statistical methods. One common approach to error adjustment is the least squares method, which minimizes the sum of the squares of the errors in the observations. This method provides a way to distribute the total error among all measurements, yielding adjusted values that improve the overall accuracy of the survey.

In the context of triangulation, error adjustment may involve recalculating angles and distances based on the least squares method to ensure that the sum of the angles in a triangle equals 180 degrees and that the calculated distances best fit the observed data. For traversing, error adjustment is used to ensure that the sum of the interior angles matches the theoretical sum based on the number of sides in the traverse polygon and that the closing error, or discrepancy between the final and initial points in a closed traverse, is minimized.

The application of triangulation and traversing in field measurements is a testament to the precision and rigor required in surveying. These methods, supported by error adjustment techniques, provide the foundation for accurate land assessment, mapping, and engineering project planning. Through the meticulous application of these techniques, surveyors can achieve the high level of accuracy necessary for the successful execution of engineering projects, ensuring that structures are built in the correct locations and to the proper specifications.

Area Computations

Area Calculations for Land Parcels

Calculating the area of land parcels is a fundamental task in surveying that requires precision and understanding of geometric principles. This section delves into the basic methods of area calculations using coordinates, triangles, and trapezoids, which are essential for accurate land assessment and mapping. These methods not only serve as the foundation for more complex surveying tasks but also ensure that engineers can confidently determine the size and boundaries of land parcels for development, legal, and environmental purposes.

The first method involves calculating the area using coordinates. This approach is particularly useful for parcels with irregular boundaries that can be broken down into a series of smaller, more manageable shapes. By plotting the parcel's corners as points in a Cartesian coordinate system, where each point has an x, y coordinate, the area can be computed using the formula for the determinant of a matrix formed by these coordinates. For a polygon with vertices $(x_1, y_1), (x_2, y_2), \ldots, (x_n, y_n)$, the area A can be calculated as:

$$A = \frac{1}{2} \left| \sum_{i=1}^{n-1} x_i y_{i+1} + x_n y_1 - \sum_{i=1}^{n-1} y_i x_{i+1} - y_n x_1 \right|$$

This formula effectively sums the areas of trapezoids formed between the polygon's sides and the x-axis, subtracting the area below the x-axis to find the total area of the polygon.

The second basic method involves using triangles. Since any polygon can be divided into triangles, calculating the area of a land parcel often involves dividing it into several triangles, calculating each area separately, and then summing these areas. The area of a triangle can be calculated using the basic formula $A = \frac{1}{2} b h$, where b is the base and h is the height.

Alternatively, for triangles where the base and height are not readily apparent, Heron's formula can be used when the lengths of all three sides, a, b, and c, are known:

$$A = \sqrt{s(s-a)(s-b)(s-c)}$$

where s is the semi-perimeter of the triangle, or $\dfrac{a+b+c}{2}$.

The third method involves trapezoids, which are particularly useful for parcels that can be approximated as or decomposed into a series of trapezoidal sections. The area of a trapezoid is given by the formula $A = \dfrac{1}{2}(b_1 + b_2)h$, where b_1 and b_2 are the lengths of the parallel sides, and h is the height, or the perpendicular distance between these sides. This method is advantageous for parcels with parallel boundary lines, allowing for straightforward decomposition into trapezoids and simplification of the area calculation process.

In practice, these methods are often used in combination, depending on the shape and complexity of the land parcel being surveyed. By breaking down a parcel into geometric shapes for which the area can be readily calculated, surveyors can accurately determine the total area of the parcel. This process not only requires a thorough understanding of geometric principles but also meticulous attention to detail in measuring and plotting the parcel's boundaries. The ability to accurately calculate land area is crucial for a wide range of engineering applications, including site planning, design, and legal documentation, underscoring the importance of mastering these basic surveying techniques.

Advanced Applications in Area Estimation

For parcels with irregular boundaries, traditional methods of area calculation may not suffice due to the complexity and non-linear edges of the land. In such cases, Simpson's rule and Geographic Information Systems (GIS) offer advanced techniques for more accurate area estimation.

Simpson's Rule is a numerical method that approximates the area under a curve by dividing the total area into a series of segments that are easier to calculate. This method is particularly useful for irregularly shaped parcels where the boundary can be approximated by a smooth curve. The formula for Simpson's rule is given by:

$$A \approx \frac{\Delta x}{3} [f(x_0) + 4f(x_1) + 2f(x_2) + 4f(x_3) + \cdots + 2f(x_{n-2}) + 4f(x_{n-1}) + f(x_n)]$$

where A is the area under the curve, Δx is the width of each segment (assumed to be equal for all segments), $f(x_i)$ is the function value at the ith point, and n is the total number of segments, which must be an even number. By applying this formula to the boundary of a land parcel, surveyors can accurately estimate its area, even if the boundary is not a simple geometric shape.

Geographic Information Systems (GIS) further enhance the capability to estimate areas with irregular boundaries by utilizing spatial data and analysis. GIS allows for the collection, storage, and analysis of geographic information, including land boundaries and topographical features. By overlaying satellite imagery or aerial photographs with vector data representing the parcel boundaries, GIS can calculate the area of irregular shapes with high precision. The process involves digitizing the boundary of the parcel into a series of points or vectors, which GIS software then uses to compute the area based on spatial analysis algorithms.

Integrating GIS for area estimation not only provides a high degree of accuracy but also offers flexibility in handling various land shapes and sizes. Additionally, GIS databases can store historical data, allowing for comparisons over time and the analysis of land use changes. This capability is invaluable for engineers and surveyors working on projects that require detailed environmental impact assessments or land use planning.

The combination of Simpson's rule and GIS technology represents a powerful toolset for surveyors and engineers. By leveraging these advanced applications, professionals can overcome the challenges associated with estimating the area of parcels with irregular boundaries, ensuring precise calculations for project planning, design, and legal documentation. This approach not only saves time but also reduces the potential for error in manual calculations, contributing to more efficient and effective surveying practices.

Earthwork and Volume Computations

Cut and Fill: Earthwork Volume Computation

The computation of earthwork volumes, particularly through **cut and fill** operations, is a critical task in the planning and execution of construction projects. This process involves the calculation of the volume of soil to be removed (cut) and the volume of soil to be added (fill) to reach a specified design grade. The accuracy of these computations directly impacts project costs, scheduling, and environmental considerations.

Cross-sections are utilized to represent the topography of the land before and after excavation or filling. These are essentially vertical slices through the terrain, depicted graphically to show elevations at specific intervals. To calculate the volume of earthwork using cross-sections, one must first draw these sections before the commencement of the project (existing ground line) and after the project completion (design grade line).

The area between the existing ground line and the design grade line on each cross-section represents the volume of soil to be cut or filled. The volume between two adjacent cross-sections can be approximated using the **Average End Area Method**. This method calculates the volume V between two sections by taking the average of their areas A_1 and A_2 and multiplying by the distance L between them:

$$V = \frac{(A_1 + A_2)}{2} \times L$$

For irregularly shaped areas, the cross-sections are divided into smaller segments, and the area of each segment is calculated using geometric formulas or numerical integration methods. The total area is then the sum of these segment areas.

Grid methods offer another approach to calculating earthwork volumes, especially useful for large or irregular sites. The site is divided into a grid of uniform squares or rectangles. The elevation at each grid point is measured before and after the earthwork operation. The difference in elevation at each grid point indicates whether a cut or fill operation is needed to achieve the design grade.

The volume of cut or fill for each grid cell is calculated by multiplying the area of the cell by the average depth of cut or fill across the cell. For a cell with an area A and an average depth of cut or fill d, the volume V is:

$$V = A \times d$$

The total volume of earthwork is the sum of the volumes of all grid cells, with cuts and fills treated separately to provide a clear picture of the total excavation and fill requirements.

Both the cross-section and grid methods require careful measurement and calculation to ensure accuracy. The choice between methods depends on the project's specific requirements, including the site's size, the complexity of the terrain, and the precision needed for the earthwork estimates.

Modern technology, such as **Geographic Information Systems (GIS)** and **Global Positioning Systems (GPS)**, can enhance the accuracy and efficiency of these calculations. These technologies allow for the rapid collection of topographic data and the automation of volume calculations, reducing the potential for human error and improving the reliability of earthwork estimates.

The computation of earthwork volumes is a foundational step in the planning and design of engineering projects. It informs budgeting, environmental impact assessments, and project scheduling, underscoring the importance of precision and accuracy in surveying and civil engineering tasks.

Advanced Volume Estimation Techniques

In the realm of surveying, particularly when addressing the complexities of earthwork and volume computations, advanced techniques such as contour integration, borrow-pit methods, and 3D modeling stand out for their precision and efficiency. These methodologies are indispensable for engineers aiming to achieve accurate volume estimates in varied terrain and project scopes.

Contour integration, a method rooted in the principles of calculus, leverages the topographical contours of the project site. By analyzing the contour lines, which represent points of equal elevation, engineers can calculate the volume of material to be excavated or filled. The process involves the creation of a series of horizontal slices through the terrain, each bounded by contour lines. The area of each slice is calculated, typically through planimetric methods or using geographic information systems (GIS), and then multiplied by the distance to the next contour to estimate the volume of each segment. The sum of these volumes provides a comprehensive estimate for the entire project. This technique is particularly effective in areas with complex topography, where traditional methods may fall short in accuracy.

The borrow-pit method, on the other hand, is commonly employed in projects requiring the excavation of material from a designated area, known as a borrow pit, to be used elsewhere within the project. This method necessitates a detailed understanding of the pit geometry and the material's compaction characteristics. The initial step involves delineating the borrow pit area and conducting a topographic survey to establish the existing ground surface. Following excavation, a second survey determines the new ground surface. The volume of excavated material is then calculated by comparing the two surfaces, taking into account factors such as swell and shrinkage, which affect the volume of soil when it is moved and compacted. This method ensures that the quantity of material transported from the borrow pit matches the project's requirements, thereby optimizing resource utilization.

3D modeling represents the forefront of volume estimation techniques, offering a dynamic and highly visual approach. Utilizing advanced software, engineers can create detailed three-dimensional representations of the project area, incorporating data from topographical surveys, satellite imagery, and drone reconnaissance. These models allow for precise manipulation and analysis, enabling engineers to simulate various scenarios, including excavation, fill, and grading operations. The ability to visualize the project in three dimensions aids in identifying potential issues and optimizing the design before any physical work begins. Moreover, 3D modeling facilitates the calculation of volumes with exceptional accuracy, as the software can precisely account for the complexities of the terrain and the specific characteristics of the materials involved.

Each of these advanced techniques—contour integration, borrow-pit methods, and 3D modeling—plays a crucial role in modern surveying practices. They provide engineers with the tools necessary to tackle the challenges of accurately estimating earthwork volumes, ensuring that projects are executed efficiently, within budget, and in compliance with environmental and regulatory standards. As technology continues to evolve, these methods will undoubtedly become even more integral to the field of surveying, pushing the boundaries of what is possible in engineering projects.

Coordinate Systems

Basics of Coordinate Systems

In the realm of surveying and civil engineering, understanding the basics of coordinate systems is fundamental for accurate mapping and project planning. Coordinate systems provide a framework for locating points on the Earth's surface or in its geometric space, facilitating the precise representation of features, structures, and boundaries in engineering projects.

The State Plane Coordinate System (SPCS) is a set of 124 geographic zones or coordinate systems designed for specific regions of the United States. Each state may be divided into one or more state plane zones, each of which uses a specific map projection suited to its shape, size, and geographic location. The SPCS allows for the use of a flat Cartesian coordinate system on a curved surface, enabling measurements in feet or meters with minimal distortion over small areas, which is ideal for engineering projects. The system is based on either the Lambert Conformal Conic projection for states that are longer in the east-west direction or the Transverse Mercator projection for states that are longer in the north-south direction. For states with panhandles or irregular shapes, the Oblique Mercator projection may be used. The primary advantage of the SPCS is its ability to provide highly accurate location data within the specific zone, minimizing errors in measurements and calculations critical to civil engineering projects.

Latitude and longitude coordinates, on the other hand, offer a global reference system for locating points on the Earth's surface. Latitude measures the angular distance of a point north or south of the Equator, while longitude measures the angular distance of a point east or west of the Prime Meridian. These measurements are expressed in degrees, minutes, and seconds. Latitude and longitude are essential for global navigation and positioning but can introduce significant distortions when used for detailed, localized mapping in engineering due to the Earth's curvature. However, they are invaluable for establishing a general location within a global context, from which more precise, localized coordinate systems can take over for detailed project work.

The Universal Transverse Mercator (UTM) system divides the world into 60 north-south zones, each 6 degrees of longitude wide. This system uses a two-dimensional Cartesian coordinate system to give locations on the surface of the Earth. It is a type of cylindrical map projection where the globe is divided into zones, and within each zone, measurements are taken in meters from a central meridian and the equator. The UTM system is particularly useful for linear and area measurements because it scales and distortion are minimized within each zone. This makes UTM coordinates highly useful for detailed mapping and engineering projects that span large distances, as it allows for the use of linear units of measurement (meters) across both small and large areas with consistent accuracy.

In mapping applications, the choice between state plane systems, latitude/longitude, and UTM often depends on the scale and geographic extent of the project. For localized projects with a need for high precision, such as property surveys, construction, and infrastructure development, the state plane system may be preferred. For projects that require a broader geographic context or span multiple countries, such as environmental studies and international logistics, latitude/longitude provides a universally recognized reference. Meanwhile, UTM coordinates offer a balance between local accuracy and broader applicability, making them suitable for regional planning, navigation, and military operations.

Transformations in Geospatial Engineering

Transformations between coordinate systems are a fundamental aspect of geospatial engineering, enabling the accurate translation of spatial data from one coordinate system to another. This process is crucial for integrating data collected using different systems, ensuring consistency and precision in mapping, analysis, and project planning. The ability to effectively convert between coordinate systems such as the State Plane Coordinate System (SPCS), Universal Transverse Mercator (UTM), and latitude/longitude coordinates is essential for civil engineers and surveyors working on a wide range of projects.

The transformation process involves mathematical equations that adjust the coordinates from one system to another, taking into account the shape of the Earth (ellipsoid or geoid), the scale, and the origin of the coordinate system. The most common transformations are datum transformations, which adjust for differences in the reference frame used to define coordinate systems, and projection transformations, which convert between planar and spherical coordinate systems.

Datum Transformations: A datum is a reference from which spatial measurements are made. In geospatial engineering, transforming data from one datum to another is necessary when integrating spatial data from different sources. This transformation adjusts for variations in the Earth's shape and the origin point of the coordinate system. For example, converting coordinates from the North American Datum of 1983 (NAD83) to the World Geodetic System of 1984 (WGS84) involves slight adjustments to account for differences in the Earth model and origin points used by each datum.

Projection Transformations: Projection transformations are required when converting between spherical (latitude and longitude) and planar coordinate systems (such as SPCS or UTM). Since the Earth is a three-dimensional object, projecting its surface onto a two-dimensional plane involves mathematical models that introduce distortions in area, shape, distance, or direction. Engineers must choose the most appropriate projection based on the project requirements, minimizing distortions for the specific area of interest. For instance, the Lambert Conformal Conic projection is suitable for east-west elongated areas, while the Transverse Mercator projection is better for north-south elongated regions.

The mathematical process of transforming coordinates typically involves the use of affine transformations for linear adjustments and more complex algorithms like the Helmert transformation for datum shifts. These transformations can be represented by equations such as:

$$x' = ax + by + c$$

$$y' = dx + ey + f$$

where x' and y' are the transformed coordinates, x and y are the original coordinates, and $a, b, c, d, e,$ and f are parameters defined by the transformation process. For more complex transformations, parameters may include rotation, scaling, and translation factors that align the two coordinate systems.

Understanding and applying these transformations accurately is vital for ensuring that spatial data aligns correctly on the Earth's surface, which is essential for any engineering project that relies on precise location data. Errors in transformation can lead to significant discrepancies in project planning and execution, potentially leading to increased costs, delays, and failures in infrastructure development.

In geospatial engineering, software tools and geographic information systems (GIS) play a crucial role in managing these transformations, offering built-in functions and algorithms to convert between coordinate systems seamlessly. However, engineers must still possess a deep understanding of the principles underlying these transformations to select the appropriate methods and verify the accuracy of the results.

The importance of transformations in geospatial engineering cannot be overstated, as they enable the integration of diverse spatial data sets into a coherent framework for analysis, design, and decision-making. By mastering the art and science of coordinate system transformations, engineers ensure the reliability and accuracy of their work, contributing to the successful completion of projects that shape our built and natural environments.

Leveling

Differential Leveling Basics

Differential leveling is a fundamental surveying technique used to determine the elevation difference between two or more points. This method involves measuring vertical distances from a known elevation point to an unknown point, providing the basis for calculating elevations across a site. The process is critical for various engineering projects, including construction, infrastructure development, and environmental studies, where precise elevation data is essential for design and planning.

The procedure for differential leveling starts with the establishment of a benchmark—a point of known or assumed elevation. From this benchmark, a level instrument, typically an optical or digital level, is positioned on a stable platform at a location where both the benchmark and the point(s) whose elevation is to be determined are visible. The level is precisely adjusted to a horizontal plane using its built-in spirit or electronic level.

Once the level is set, a graduated rod or staff is placed vertically on the benchmark, and the height at which the horizontal crosshair intersects the rod is recorded. This measurement is known as the backsight (BS). The backsight is a reading taken on a point of known elevation to establish the line of sight for the level. The instrument height (IH) is then calculated by adding the backsight reading to the benchmark elevation.

Next, the rod is moved to the point of unknown elevation, and a similar reading is taken. This reading is called the foresight (FS). The foresight is a reading taken on a point of unknown elevation to determine the difference in height from the level. The elevation of the unknown point is calculated by subtracting the foresight from the instrument height.

The fundamental equation for differential leveling is given by:

$$\text{Elevation of point B} = \text{Elevation of point A} + BS - FS$$

where point A is the benchmark or any point of known elevation, and point B is the point of unknown elevation.

Accuracy in differential leveling is paramount and is influenced by several factors, including the type of leveling instrument, the precision of the rod, atmospheric conditions, and the skill of the surveyor. To mitigate errors, several practices are recommended:

1. **Balancing the Backsight and Foresight Distances**: Keeping the distance between the level and the backsight point approximately equal to the distance between the level and the foresight point helps cancel out certain types of errors, particularly those due to earth curvature and refraction.

2. **Using High-Quality Equipment**: Modern optical or digital levels with automatic compensators greatly reduce the need for manual adjustments and can improve the accuracy of measurements.

3. **Environmental Considerations**: Atmospheric conditions such as temperature, wind, and light conditions can affect the accuracy of leveling. Performing surveys during mild weather conditions, preferably in the morning or late afternoon to avoid heat waves, can enhance measurement precision.

4. **Regular Calibration and Maintenance**: Ensuring that the leveling instrument and rod are regularly calibrated and maintained is crucial for accurate measurements. Even minor damages or misalignments can introduce significant errors.

5. **Skill and Experience of the Surveyor**: The proficiency of the surveyor in setting up the equipment, taking precise readings, and applying correction factors plays a critical role in the accuracy of differential leveling.

By adhering to these practices, surveyors can minimize errors in differential leveling, providing reliable data for engineering projects. The meticulous application of differential leveling techniques is essential for the successful execution of projects that depend on accurate elevation and grading plans, underscoring the importance of this method in the field of civil engineering and surveying.

Grade Computations in Topographic Surveying

Grade computations in topographic surveying are essential for the design and construction of various engineering projects, including roads, bridges, and drainage systems. These computations involve determining the slope or incline of the ground, which is crucial for ensuring proper drainage, stability, and functionality of structures. The percent grade is a common measure used to express the steepness of a slope, defined as the vertical rise divided by the horizontal run, multiplied by 100. Mathematically, the percent grade (G) can be expressed as:

$$G = \left(\frac{\Delta h}{\Delta d}\right) \times 100$$

where Δh is the change in elevation (vertical rise) and Δd is the horizontal distance (horizontal run) over which the change occurs. This formula is pivotal in topographic surveying as it quantifies the slope in terms that are easily understood and applied in engineering designs.

Slope staking is another critical process in topographic surveying, providing a physical representation of grade changes on a construction site. It involves placing stakes at specific locations to guide the construction of designed slopes. These stakes, marked with the required cut or fill information, indicate to construction personnel where and how much soil needs to be removed or added to achieve the desired grade. The process begins with surveyors calculating the positions and elevations of the stakes based on the design plans. They then transfer these calculations to the field, placing stakes at strategic points along the proposed construction area. Each stake is typically labeled with information indicating whether the area needs to be cut (excavated) or filled, and by how much, to reach the design grade. This method ensures that the constructed slopes meet the project's specifications for safety, function, and aesthetics.

Contour generation is a technique used to represent the topography of a land surface on a map. Contours are imaginary lines that connect points of equal elevation above a reference level, usually mean sea level. These lines provide a visual and quantitative understanding of the terrain's shape, slope, and elevation change, which is invaluable in planning and designing engineering projects. The process of generating contours involves collecting elevation data points across the survey area using methods such as differential leveling, GPS surveys, or aerial photogrammetry. This data is then analyzed and interpolated to draw contour lines at specific elevation intervals, depending on the scale and detail required for the project. The spacing of contour lines indicates the slope of the terrain; closely spaced contours represent steep slopes, while widely spaced contours indicate gentle slopes. Understanding how to interpret and utilize contour lines is fundamental for engineers and surveyors in assessing the feasibility of construction projects, estimating costs, and identifying potential challenges related to the terrain.

In summary, grade computations, slope staking, and contour generation are interconnected processes that play a vital role in topographic surveying. They provide engineers and surveyors with the tools to accurately describe, design, and modify the earth's surface for various construction projects. Mastery of these techniques enables the successful execution of projects that are safe, functional, and harmonious with the natural landscape, ultimately contributing to the advancement of civil engineering and surveying practices.

Chapter 10: Water Resources & Environmental Engineering

Basic Hydrology

Hydrologic Processes in Water Resource Systems

Hydrologic processes play a pivotal role in water resource systems, encompassing the dynamics of **rainfall**, **infiltration**, **runoff**, and the characteristics of **watersheds**. Understanding these processes is essential for the effective management and conservation of water resources.

Rainfall is the primary source of water for most watersheds. It varies significantly across different geographical locations and seasons. Quantifying rainfall involves measuring its intensity, duration, and frequency, which are critical for hydrologic analysis and design. The intensity of rainfall directly influences the infiltration and runoff processes, affecting the water availability in a watershed.

Infiltration refers to the process by which water on the ground surface enters the soil. It is governed by several factors, including soil type, vegetation cover, soil moisture content, and the presence of impervious surfaces. The rate of infiltration decreases as the soil becomes saturated. Engineers use the infiltration rate to estimate the volume of water that percolates into the ground, which is crucial for groundwater recharge assessments and designing stormwater management systems.

Runoff occurs when the rate of rainfall exceeds the ground's capacity to absorb water, leading to the flow of water over the land's surface. This process is influenced by the watershed's topography, land use, soil type, and vegetation. Runoff is a key component in hydrologic studies, as it affects flood peak discharges, river flow rates, and sediment transport. Calculating runoff involves understanding the watershed's response to precipitation events, which can be modeled using various hydrologic models to predict flow rates and volumes.

The **watershed** or drainage basin is an area of land that drains all the streams and rainfall to a common outlet. Watershed characteristics, including area, shape, slope, soil type, and land use, significantly influence the hydrologic response of the watershed. The delineation of watersheds is critical for managing water resources, designing flood control measures, and assessing environmental impacts.

Hydrologists and engineers analyze these hydrologic processes using empirical formulas and hydrologic models. For instance, the **Curve Number (CN)** method developed by the **Soil Conservation Service (SCS)** is widely used to estimate direct runoff from rainfall events. Similarly, the **Rational Method** is employed for calculating peak discharge rates for small urban catchments. Advanced hydrologic modeling software, such as **HEC-HMS** and **SWMM**, offer sophisticated tools for simulating the hydrologic cycle in watersheds, enabling the design of efficient water resource management and flood mitigation strategies.

Hydrology Applications

Hydrographs are essential tools in hydrology for representing the flow rate of water through a river or stream over time, typically in response to a precipitation event. They are characterized by a rising limb, a peak flow, and a falling limb, providing valuable insights into the watershed's response to rainfall. The peak flow, or the highest point on the hydrograph, is of particular interest as it indicates the maximum rate of flow during an event, critical for designing flood control measures and infrastructure resilience.

Streamflow measurements are conducted using various methods, including direct measurements with flow meters in smaller channels and indirect methods for larger rivers, such as using rating curves that correlate water levels (stage) to flow rates. These measurements are fundamental for calibrating hydrologic models and validating their predictions against observed data.

Estimating peak flows is a pivotal aspect of hydrologic analysis, especially in the design of hydraulic structures like culverts, bridges, and stormwater management systems. Engineers employ several models to predict peak flows, including empirical formulas, statistical analyses, and physically based models. The Rational Method is a widely used empirical formula for estimating peak discharge from small catchment areas, given by:

$$Q = \frac{CIA}{360}$$

where Q is the peak discharge (cfs), C is the runoff coefficient, I is the rainfall intensity (in/hr) for a duration equal to the time of concentration of the watershed, and A is the catchment area (acres). The runoff coefficient C reflects the proportion of rainfall that will become runoff, varying with land use, soil type, and slope.

For larger watersheds or more complex analyses, hydrologists may use statistical methods like frequency analysis to estimate peak flows based on historical flow records. This approach involves fitting a probability distribution to observed peak flows and estimating the flow for a specific return period, such as a 100-year flood.

Physically based hydrologic models, such as the Hydrologic Engineering Center's Hydrologic Modeling System (HEC-HMS), simulate the rainfall-runoff processes in a watershed. These models can account for various hydrologic phenomena, including interception, infiltration, surface runoff, and baseflow, providing a comprehensive tool for estimating peak flows under different scenarios.

In applying these models, engineers must consider the accuracy of input data, including precipitation patterns, watershed characteristics, and land use changes, which can significantly impact model predictions. Calibration and validation of models against observed data are crucial steps to ensure their reliability for decision-making in water resources planning and management.

By leveraging hydrographs, streamflow measurements, and peak flow estimation models, engineers can design and manage water resources systems more effectively, ensuring resilience against flooding and optimizing the allocation of water resources for human and environmental needs.

Basic Hydraulics

Flow Principles and Applications

Understanding the principles of fluid flow is crucial for civil engineers, especially when dealing with water resources and environmental engineering projects. Two fundamental concepts in this realm are the Bernoulli theorem and the continuity equation, both of which have wide-ranging applications in both closed and open-channel flow scenarios.

The Bernoulli theorem, derived from the principle of energy conservation, states that for an incompressible, frictionless fluid, the total mechanical energy along a streamline remains constant. Mathematically, it can be expressed as:

$$P + \frac{1}{2}\rho v^2 + \rho g h = \text{constant}$$

where P represents the pressure energy per unit volume, $\frac{1}{2}\rho v^2$ the kinetic energy per unit volume, and $\rho g h$ the potential energy per unit volume. Here, ρ is the fluid density, v is the fluid velocity, g is the acceleration due to gravity, and h is the elevation head. This theorem is instrumental in analyzing fluid behavior in various engineering applications, such as the flow through pipes, around airfoils, and through channels. For instance, it allows engineers to calculate the pressure distribution along a pipeline or to determine the velocity of water exiting a spillway.

The continuity equation, on the other hand, is a statement of mass conservation in fluid flow. For an incompressible fluid, it implies that the mass flow rate through a pipe or channel is constant across any cross-section. The equation can be written as:

$$A_1 v_1 = A_2 v_2$$

where A_1 and A_2 are the cross-sectional areas at points 1 and 2, and v_1 and v_2 are the fluid velocities at these points. This equation is fundamental in ensuring that fluid flow calculations account for changes in velocity and area, such as when a river narrows or widens, or when a pipe changes diameter. In practical terms, it helps engineers design systems that maintain desired flow rates and pressures, ensuring, for example, that water distribution networks can adequately supply a community without overloading or underutilizing parts of the system.

In closed-channel flow, such as in pipes or conduits, these principles enable the analysis and design of efficient systems for transporting water or other fluids with minimal energy loss. Engineers can calculate the necessary pipe diameters, pump specifications, and system layout to achieve desired flow characteristics, taking into account the effects of friction and minor losses which are not considered in the idealized forms of Bernoulli's theorem and the continuity equation.

For open-channel flow, such as in rivers, canals, and spillways, these principles are adapted to account for the free surface of the fluid. The Bernoulli theorem is applied along a streamline that is parallel to the free surface, and the continuity equation helps in understanding how the flow depth changes with channel width and slope, which is critical for flood management, irrigation, and environmental protection projects.

By applying the Bernoulli theorem and continuity equation, engineers can predict how water will behave under various conditions, design structures that control or utilize flow effectively, and solve problems related to water distribution, wastewater treatment, and environmental management. These principles are foundational to the field of water resources and environmental engineering, providing the tools necessary to design systems that are both efficient and sustainable, ensuring the careful management of one of our most precious resources: water.

Manning Equation and Channel Design

The Manning equation, a cornerstone in hydraulic engineering, provides a method for calculating the velocity of flow in a channel or a pipe. It is an empirical formula that incorporates the channel's physical characteristics and the nature of the fluid. The equation is expressed as:

$$v = \frac{1}{n} R^{2/3} S^{1/2}$$

where v is the velocity of the fluid (ft/s), n is the Manning roughness coefficient, R is the hydraulic radius (ft), which is the area of the cross-section of flow divided by the wetted perimeter, and S is the slope of the energy grade line or the channel bed slope (ft/ft).

The Manning roughness coefficient, n, is a dimensionless quantity that characterizes the internal surface roughness of the channel. It varies with the type of channel material and the degree of irregularity and vegetation. For instance, a smooth concrete surface will have a lower n value, indicating less resistance to flow, whereas a rough, vegetated channel bed will have a higher n value.

The hydraulic radius, R, is a measure of the efficiency of the channel shape in conveying water. A larger hydraulic radius suggests a more efficient cross-section, with less resistance to flow per unit of area. It is crucial in the design of open channels to maximize the hydraulic radius by optimizing the channel's cross-sectional shape.

The slope, S, directly influences the gravitational force component acting on the fluid, driving its movement. A steeper slope results in a higher velocity, assuming all other factors remain constant.

In designing efficient open channels, engineers must carefully consider these variables. The goal is to achieve a balance between the hydraulic efficiency, construction and maintenance costs, and environmental and land use constraints. The Manning equation serves as a guide in this optimization process, allowing for the prediction of flow velocities and the assessment of different channel designs under varying flow conditions.

For practical applications, the Manning equation is used to determine the required channel dimensions to convey a specific flow rate at a given slope, taking into account the expected roughness of the channel's boundary surfaces. This involves selecting an appropriate value for n based on the channel's construction material and surface condition, calculating the hydraulic radius for various cross-sectional shapes, and evaluating the impact of the slope on the flow velocity.

In open channel design, attention must also be given to the flow profile, which describes the depth and velocity of flow at different sections along the channel. The flow profile is influenced by the channel's slope, roughness, and shape, as well as by upstream and downstream boundary conditions. Engineers use flow profiles to identify potential problems such as backwater effects, which occur when downstream conditions cause the water to back up, and critical flow conditions, where the channel transitions between subcritical and supercritical flow states.

The design of efficient open channels is a complex task that requires a thorough understanding of fluid dynamics, channel hydraulics, and environmental considerations. By applying the Manning equation and considering the flow profiles, engineers can design channels that effectively manage water resources, protect against flooding, and meet the needs of agricultural, industrial, and urban systems.

Pumps

Pump Types and Performance Basics

In the realm of water resources and environmental engineering, understanding the fundamentals of pump operation, including the types of pumps and their performance curves, is essential for the design and analysis of fluid transport systems. Pumps are mechanical devices that convert electrical or mechanical energy into hydraulic energy, facilitating the movement of fluids from one location to another. The two primary categories of pumps encountered in engineering practices are centrifugal pumps and positive displacement pumps, each with distinct operational mechanisms and applications.

Centrifugal pumps operate on the principle of imparting velocity to the fluid through a rotating impeller, which then converts this velocity into flow. Inside a centrifugal pump, the impeller's rotation generates a centrifugal force that pushes the fluid outward from the pump inlet to its outlet, through the pump casing. The energy transfer in centrifugal pumps occurs through dynamic interactions between the pump and the fluid; the velocity imparted by the impeller is transformed into pressure when the fluid exits the impeller into the pump casing. This transformation is described by the Bernoulli equation, which relates the increase in pressure to the decrease in fluid velocity, conserving the total mechanical energy. The performance of centrifugal pumps is significantly influenced by the flow rate, with the head (pressure) generated by the pump decreasing as the flow rate increases. This relationship is graphically represented in the pump's performance curve, which plots the pump head against the flow rate, illustrating how the pump operates under different conditions.

Positive displacement pumps, in contrast, operate by trapping a fixed amount of fluid and forcing (displacing) that trapped volume into the discharge pipe. Unlike centrifugal pumps, positive displacement pumps can theoretically provide the same flow at a given speed no matter what the discharge pressure is. Therefore, these pumps are suitable for applications requiring a consistent flow rate regardless of the pressure conditions. Positive displacement pumps are further categorized into rotary and reciprocating types, with mechanisms varying from gears, screws, lobes for rotary pumps to pistons, diaphragms, and plungers for reciprocating pumps. The performance of positive displacement pumps is characterized by a nearly constant flow rate over a wide range of discharge pressures, depicted in their performance curves. These curves are typically flatter than those of centrifugal pumps, indicating less variation in flow with changes in pressure.

The pump performance curve is a critical tool for engineers, providing vital information for selecting the appropriate pump for a specific application. The curve not only shows the relationship between the head and the flow rate but also includes efficiency curves, power consumption, and NPSH required (Net Positive Suction Head) to avoid cavitation within the pump. Understanding these curves allows engineers to match the pump to the system's requirements, ensuring optimal performance while minimizing energy consumption and wear.

When integrating pumps into water resources and environmental engineering projects, engineers must consider various factors, including the type of fluid to be pumped, the required flow rate and pressure, the efficiency of the pump, and the total cost of ownership, which includes initial costs, operating costs, maintenance, and downtime. The choice between a centrifugal and a positive displacement pump depends on the specific demands of the application, with centrifugal pumps generally preferred for high flow, low pressure applications, and positive displacement pumps favored for low flow, high pressure scenarios.

The selection and application of pumps in engineering projects require a thorough understanding of the different types of pumps, their operational principles, and performance characteristics. By analyzing pump performance curves and considering the system's hydraulic requirements, engineers can ensure the efficient and reliable operation of pumps in water resources and environmental engineering systems, contributing to the sustainability and effectiveness of water management practices.

System Integration in Hydraulic Systems

In the realm of hydraulic systems, the integration of pumps is a critical factor that influences the overall efficiency and functionality of water resources and environmental engineering projects. The selection of the appropriate pump type is paramount and hinges on a comprehensive understanding of the system requirements, including flow rate, head, fluid characteristics, and the specific application at hand. The process begins with an assessment of the required flow rate (Q), which is typically measured in gallons per minute (GPM) or cubic meters per hour (m³/h), and the total dynamic head (TDH), which encompasses both the static lift and friction losses within the system and is measured in feet (ft) or meters (m).

The selection criteria extend to evaluating pump efficiency, which is a measure of the pump's ability to convert mechanical energy into hydraulic energy. The efficiency of a pump (η) is calculated as the ratio of the hydraulic power output to the mechanical power input, expressed as a percentage. High-efficiency pumps are sought after for their ability to minimize energy consumption and operational costs over the pump's lifecycle. It is crucial to match the pump's performance curve with the system's requirements to ensure optimal efficiency under operating conditions.

Head loss is another critical consideration in system integration. It occurs due to friction within pipes, fittings, valves, and other components, and it affects the flow rate and pressure within the system. The Darcy-Weisbach equation, $\Delta h = f \cdot \left(\frac{L}{D}\right) \cdot \left(\frac{v^2}{2g}\right)$, where Δh is the head loss, f is the friction factor, L is the length of the pipe, D is the diameter of the pipe, v is the velocity of the fluid, and g is the acceleration due to gravity, is commonly used to calculate head loss in pipes. Understanding and minimizing head loss through careful design and selection of components is essential for maintaining the efficiency of the hydraulic system.

Efficiency considerations extend beyond the selection phase into the operational and maintenance practices. Regular monitoring of pump performance and system parameters can identify inefficiencies and potential issues before they escalate into significant problems. Troubleshooting in hydraulic systems often involves analyzing symptoms such as reduced flow rate, increased power consumption, abnormal noises, or overheating, which can indicate issues like clogging, wear, improper installation, or mechanical failure. Effective troubleshooting requires a systematic approach to isolate the cause of the problem, which may involve checking for blockages, ensuring the pump is primed, verifying the electrical connections, and inspecting the pump and system components for wear or damage.

The integration of pumps into hydraulic systems is a multifaceted process that demands a thorough understanding of both the system requirements and the characteristics of the pumps. Selecting the right pump, minimizing head loss, maintaining high efficiency, and effective troubleshooting are all critical for the successful operation of hydraulic systems in water resources and environmental engineering applications.

Water Distribution Systems

System Components in Water Distribution

In the realm of water distribution systems, understanding the integration and functionality of system components such as pipe networks, reservoirs, valves, and pumps is paramount for engineers aiming to ensure efficient water delivery and management. These components form the backbone of water distribution systems, each playing a critical role in the transportation, regulation, storage, and control of water flow to meet various demands and ensure reliability under different operational conditions.

Pipe networks are the primary conduits through which water is distributed from treatment plants to consumers. The design of these networks involves the selection of appropriate pipe sizes, materials, and layouts to minimize head loss and ensure adequate pressure at all service points. The hydraulic design of pipe networks relies on principles such as the Hazen-Williams equation, $\Delta P = 4.52 Q^{1.85}/C^{1.85} D^{4.87}$, where ΔP represents the pressure loss in psi, Q is the flow rate in gallons per minute, C is the Hazen-Williams roughness coefficient, and D is the pipe diameter in inches. This equation helps in determining the pressure drops across the network, ensuring that all areas receive water at sufficient pressures.

Reservoirs serve as storage units within water distribution systems, providing a buffer to accommodate fluctuating demands and maintaining system pressure. They are strategically located to utilize gravity for pressure regulation, reducing the reliance on pumps and thereby saving energy. The volume and elevation of reservoirs are critical design parameters that are calculated based on daily consumption rates, emergency storage, and fire flow requirements. The effective volume of a reservoir, V, can be estimated by considering the demand variation, D, and the time period, T, over which the storage is to be provided, using the formula $V = D \times T$.

Valves are essential for controlling the flow of water within the distribution network, allowing for isolation, regulation, and protection of the system. Types of valves commonly used include gate valves for isolation, pressure-reducing valves to control pressure, check valves to prevent backflow, and air release valves to remove entrapped air from pipes. The selection and placement of valves are critical for operational efficiency and maintenance, ensuring that parts of the network can be isolated for repairs without disrupting the entire system.

Pumps are utilized to move water from lower to higher elevations or across long distances where gravity flow is insufficient. The selection of pumps is based on the required flow rate and head, with centrifugal pumps being widely used due to their efficiency and capacity to handle large volumes of water. The pump performance curve, which plots the relationship between the flow rate and the head, guides the selection process, ensuring that the pump operates at or near its best efficiency point (BEP). The operating point of a pump in a system is determined by the intersection of the pump curve and the system curve, which represents the relationship between the flow rate and the head loss in the system.

The design and operation of water distribution systems hinge on a thorough understanding of the interplay between pipe networks, reservoirs, valves, and pumps. Engineers must meticulously calculate and optimize each component's specifications to achieve a balance between efficiency, reliability, and cost-effectiveness, ensuring that water distribution systems operate seamlessly to meet the demands of the communities they serve.

Hydraulic Analysis: Pressure, Demand, Optimization

Hydraulic analysis in water distribution systems is pivotal for ensuring efficient operation and management of water resources. This analysis encompasses **pressure calculations**, **water demand assessment**, and **network optimization** to meet the varying needs of a community while minimizing costs and maximizing system reliability and performance.

Pressure Calculations in water distribution systems are fundamental for maintaining water flow and serviceability across all points of the network. The Hazen-Williams formula, given by $H_f = 10.67 \cdot L \cdot Q^{1.85} / (C^{1.85} \cdot D^{4.8655})$, where H_f is the head loss in feet, L is the length of the pipe in feet, Q is the flow rate in gallons per minute (GPM), C is the Hazen-Williams roughness coefficient, and D is the diameter of the pipe in inches, is widely used for this purpose. Accurate pressure calculations ensure that the system can deliver water effectively to all users, taking into account the frictional losses that occur as water moves through pipes.

Water Demand Assessment involves estimating the quantity of water required by the community at different times of the day, month, or year. This assessment must consider not only the average daily demand but also peak demand periods, which can significantly strain the distribution system. Engineers use historical consumption data, demographic trends, and specific community needs to model water demand, often employing software tools that can simulate various demand scenarios. Understanding these patterns is crucial for designing a system that can accommodate fluctuations without compromising water pressure or availability.

Network Optimization focuses on configuring the water distribution system to operate efficiently under all demand conditions. This involves the strategic placement of pipes, pumps, and reservoirs; selecting appropriate pipe diameters and materials to minimize head losses; and implementing advanced control systems for pumps and valves to adjust the flow dynamically based on real-time demand. Optimization techniques may include the use of genetic algorithms, linear programming, or hydraulic simulation models to identify the most cost-effective solutions that meet all regulatory and service quality standards.

The goal of hydraulic analysis is to achieve a balance between **cost**, **efficiency**, and **reliability**. Engineers must consider the initial construction costs, ongoing operational and maintenance expenses, and the potential for future expansions or modifications to the system. By accurately calculating pressures, assessing water demand, and optimizing the network, engineers can design water distribution systems that are robust, flexible, and capable of meeting the needs of the community both now and in the future.

The integration of **geographic information systems (GIS)** and **hydraulic modeling software** has significantly enhanced the ability to perform detailed hydraulic analyses. These tools allow for the visualization of the entire water distribution network, including pipes, valves, pumps, and reservoirs, in relation to the geographic layout of the service area. They enable engineers to simulate various operational scenarios, such as the impact of a new subdivision on water demand or the effects of a pump failure on system pressures. This level of analysis is essential for identifying potential problems, optimizing system performance, and planning for future growth or changes in water use patterns.

By employing rigorous hydraulic analysis techniques, engineers ensure that water distribution systems are designed and operated to meet the highest standards of efficiency, reliability, and sustainability. This not only supports the immediate needs of the community but also lays the foundation for accommodating future demands, technological advancements, and environmental considerations.

Flood Control

Structural Methods for Flood Management

Dams, levees, spillways, and reservoirs are critical structural methods employed in flood management to protect communities, manage water resources, and mitigate the adverse effects of excessive water flow. These structures are designed based on hydraulic and civil engineering principles to control or redirect the natural flow of water, thereby preventing flood damage.

Dams are barriers constructed across rivers or streams to control water flow and accumulation. They serve multiple purposes, including flood control, irrigation, water supply, and hydroelectric power generation. The design of a dam considers factors such as the catchment area, sediment load, peak inflow, and the potential maximum flood. The **Hydrologic Engineering Center's River Analysis System (HEC-RAS)**, for example, is often used to model water flow and evaluate dam performance under various scenarios. The structural integrity of dams is critical, requiring detailed analysis of the foundation, materials, and potential for seismic activity. The core equation governing the design of a dam for flood control is the **mass balance equation**, represented as $Q_{in} - Q_{out} = \Delta S$, where Q_{in} and Q_{out} are the inflow and outflow rates, respectively, and ΔS is the change in storage volume.

Levees are embankments built alongside rivers to prevent overflow and protect the land behind them from flooding. The design of levees involves determining the appropriate height and width to withstand expected flood levels, incorporating factors such as soil type, permeability, and the presence of wave action. The **Froehlich's formula** for crest height calculation, $H = H_w + W_s + F_a$, where H is the total levee height, H_w is the water height, W_s is the wave setup, and F_a is the freeboard allowance, is a critical consideration in levee design to ensure adequate protection.

Spillways are structures constructed to provide a controlled release of flows from a dam into a downstream area, typically a river. They are essential for preventing water from overtopping the dam, which could lead to dam failure and catastrophic flooding. The design of spillways involves hydraulic calculations to ensure they can handle the maximum probable flood (MPF). The **Weir equation**, $Q = C_w L H^{1.5}$, where Q is the flow rate, C_w is the weir coefficient, L is the length of the weir, and H is the head over the weir, is commonly used to size spillways appropriately.

Reservoirs, created by dams, play a significant role in flood control by storing excess water during periods of high inflow and releasing it slowly over time. The capacity of a reservoir to mitigate flooding is determined by its **storage curve**, which relates the volume of water stored to the water surface elevation. The operation of reservoirs for flood control involves strategic decisions about when to store and release water, often based on forecasts of incoming storms and historical flow data.

The integration of these structural methods into a comprehensive flood management strategy requires a deep understanding of hydrologic and hydraulic engineering principles. Engineers must consider the interdependencies of these structures and their impact on the watershed. Advanced modeling tools, such as **GIS-based hydrologic models** and **computational fluid dynamics (CFD)** software, are invaluable for simulating flood scenarios and optimizing the design and operation of dams, levees, spillways, and reservoirs. These tools enable engineers to predict the behavior of water systems under various conditions, assess potential risks, and implement flood management strategies that protect communities while minimizing environmental impact.

Flood Routing and Hydraulic Models

Flood routing is a critical aspect of water resources and environmental engineering, particularly in the context of flood control. It involves the prediction of how a flood wave moves through a river or canal system, which is essential for the design and operation of effective flood management strategies. This section delves into the principles of flood wave propagation, the role of detention basins, and the application of hydraulic models in flood routing.

Flood Wave Propagation refers to the movement of a flood wave down a river or through a reservoir. The speed and shape of the wave are influenced by the channel's characteristics, including its slope, roughness, and cross-sectional shape. The **Saint-Venant equations** provide the foundation for understanding flood wave propagation. These equations are a set of partial differential equations that describe the conservation of mass and momentum in open channel flow. The equations can be expressed as:

$$\frac{\partial A}{\partial t} + \frac{\partial Q}{\partial x} = 0$$

$$\frac{\partial Q}{\partial t} + \frac{\partial}{\partial x}\left(\frac{Q^2}{A}\right) + gA\frac{\partial y}{\partial x} + gA(S_f - S_0) = 0$$

where A is the cross-sectional area of flow, Q is the flow rate, t is time, x is distance along the channel, y is the depth of flow, g is the acceleration due to gravity, S_f is the friction slope, and S_0 is the bed slope.

Detention Basins play a pivotal role in managing flood waves by temporarily storing floodwaters and releasing them at a controlled rate. The design of a detention basin is based on the volume of water it needs to detain and the rate at which water can be safely released downstream. The primary objective is to mitigate peak flows and delay flood wave propagation to reduce the risk of flooding downstream. The storage capacity, V, required for a detention basin can be estimated using the equation:

$$V = \int_{t_1}^{t_2} (Q_{in} - Q_{out}) dt$$

where Q_{in} is the inflow rate, Q_{out} is the outflow rate, and t_1 and t_2 are the start and end times of the flood event, respectively.

Hydraulic Models are indispensable tools for simulating flood routing processes. These models can range from simple empirical formulas to complex numerical models that solve the Saint-Venant equations. Hydraulic models enable engineers to predict how different scenarios, such as varying rainfall events or changes in land use, will affect flood wave propagation and the performance of flood control structures like detention basins. Commonly used hydraulic models for flood routing include HEC-RAS (Hydrologic Engineering Center's River Analysis System), which can perform one-dimensional steady and unsteady flow calculations, and SWMM (Storm Water Management Model), which is used for urban drainage systems including runoff, conduits, and detention basins.

The application of these principles and tools in flood routing requires a thorough understanding of hydrology, hydraulics, and computational modeling. By accurately predicting flood wave propagation and designing detention basins to manage peak flows, engineers can develop effective flood control measures that protect communities and infrastructure from the devastating impacts of flooding.

Stormwater Management

Management Systems: Stormwater and Drainage Design

Detention and retention basins are critical components in stormwater management systems, designed to mitigate the adverse effects of urban runoff and prevent flooding. Detention basins, also known as dry ponds, temporarily store stormwater and release it at a controlled rate to downstream water bodies, reducing peak flow rates and alleviating stress on storm sewer systems. The design of a detention basin involves calculating the volume of runoff using methods such as the Rational Method, where the runoff volume V can be estimated by the equation $V = C \times I \times A$, where C is the runoff coefficient, I is the rainfall intensity, and A is the drainage area.

Retention basins, or wet ponds, serve a dual purpose: they detain stormwater and facilitate the removal of pollutants through sedimentation and biological uptake. Unlike detention basins, retention basins are designed to maintain a permanent pool of water, creating a habitat for wildlife and contributing to the aesthetic value of the area. The design of retention basins requires careful consideration of the water balance, including inputs from rainfall and runoff, and losses through evaporation, infiltration, and outflow. The sizing of retention basins can be guided by the volume control method, which aims to capture and treat a specific volume of runoff, typically the first flush, which is often considered the most polluted.

Stormwater routing through these basins is governed by the principles of hydrology and hydraulics. The design must ensure that the outflow does not exceed the capacity of downstream watercourses or storm sewer systems. This is typically achieved by incorporating outlet structures that regulate the flow, such as orifices, weirs, and outlet pipes. The design of these structures requires the application of Bernoulli's equation and Manning's equation to calculate flow rates and velocities, ensuring that the system operates within its designed parameters under varying storm conditions.

Drainage design is an integral part of stormwater management, focusing on efficiently conveying runoff from urban areas to detention or retention basins, or directly to receiving waters. The design of drainage systems involves the layout of storm sewers, culverts, and open channels, ensuring adequate capacity to handle design storm events. Hydraulic calculations for these components are based on Manning's equation, where the flow rate Q is determined by the equation $Q = \frac{1}{n} A R^{2/3} S^{1/2}$, with n being the Manning's roughness coefficient, A the cross-sectional area of flow, R the hydraulic radius, and S the slope of the energy grade line.

Incorporating green infrastructure practices, such as bioswales and permeable pavements, into the design of stormwater management systems can further enhance their effectiveness. These practices promote infiltration, evapotranspiration, and the natural treatment of stormwater, reducing the volume of runoff and improving water quality before it reaches detention or retention basins.

The design of stormwater management systems, including detention and retention basins, stormwater routing, and drainage design, requires a multidisciplinary approach, integrating principles of civil engineering, hydrology, and environmental science. By carefully considering these elements, engineers can develop effective stormwater management solutions that protect communities from flooding, enhance water quality, and contribute to the sustainability of water resources.

Water Quality and Pollutant Removal

Improving water quality through pollutant removal and infiltration practices is a critical aspect of stormwater management, aiming to address both quantity and quality challenges associated with urban runoff. The primary goal is to minimize the environmental impact of stormwater discharges into natural water bodies, thereby protecting ecosystems and human health. This section delves into the mechanisms and strategies employed to enhance water quality, focusing on the removal of pollutants and the promotion of infiltration.

Pollutant removal from stormwater is achieved through a combination of physical, chemical, and biological processes, each targeting specific contaminants. Sedimentation is a fundamental physical process where gravity assists in settling particulate matter. This is particularly effective in detention and retention basins, where the reduced flow velocity allows larger particles to settle. However, finer particles and dissolved pollutants require additional treatment methods. Filtration through vegetated swales or biofilters can remove finer sediments and associated pollutants by percolating stormwater through media layers, typically sand, gravel, and organic matter, which trap particles and adsorb contaminants.

Chemical treatment involves the use of sorbents, such as activated carbon, to remove dissolved pollutants through adsorption. This method is effective for a wide range of contaminants, including heavy metals and hydrocarbons. Additionally, constructed wetlands harness both physical and chemical processes, along with biological mechanisms, to improve water quality.

The vegetation in wetlands uptakes nutrients and heavy metals, while microbial communities degrade organic pollutants, transforming contaminants into less harmful substances.

Bioretention cells, another key element in stormwater management, combine physical filtration with biological and chemical processes. These systems are designed to mimic the natural hydrological cycle, enhancing infiltration and treating stormwater through multiple layers of soil, vegetation, and drainage media. The vegetation plays a crucial role in pollutant removal, not only through direct uptake but also by fostering a conducive environment for microbial degradation of organic pollutants.

Infiltration practices are integral to managing stormwater quality by reducing runoff volume and promoting groundwater recharge. Permeable pavements, green roofs, and infiltration trenches are designed to allow stormwater to percolate through the surface, filtering out pollutants and recharging aquifers. These practices not only mitigate surface runoff and reduce the burden on stormwater infrastructure but also enhance water quality by filtering pollutants through soil layers. The effectiveness of infiltration practices is highly dependent on soil type, permeability, and the presence of a suitable underdrain system to prevent saturation and ensure efficient operation.

The design and implementation of stormwater management systems that focus on water quality require a thorough understanding of the local hydrology, soil characteristics, and pollutant loadings. It is essential to tailor the combination of pollutant removal and infiltration practices to the specific conditions of the site to achieve optimal performance. Regular maintenance of these systems is also crucial to sustain their effectiveness over time, including sediment removal, vegetation management, and inspection of structural components.

By integrating these practices into urban planning and development, engineers can significantly contribute to the sustainability of water resources and the resilience of communities against water-related challenges. The adoption of green infrastructure and innovative stormwater management solutions not only addresses regulatory requirements but also enhances the livability of urban environments, promoting biodiversity and providing aesthetic and recreational benefits.

Collection Systems

Stormwater Collection Network Design

Designing an efficient stormwater collection network is crucial for managing runoff in urban areas, preventing flooding, and protecting water quality. The network typically comprises interconnected components including **catch basins**, **culverts**, and **storm sewers** that collect, convey, and discharge stormwater to a designated outlet such as a detention basin, water treatment facility, or natural water body. This section delves into the design considerations and functionalities of catch basins and culverts within the stormwater collection system.

Catch Basins are critical elements in the stormwater collection network, designed to capture runoff from streets, parking lots, and other impervious surfaces. They are strategically placed at low points and along curbs to maximize collection efficiency. The primary function of a catch basin is to collect sediment, debris, and pollutants carried by stormwater, preventing them from entering the storm sewer system or natural water bodies. Design considerations for catch basins include:

- **Size and Capacity**: The size of the catch basin is determined based on the expected runoff volume, with additional capacity for sediment accumulation. It is essential to ensure that the catch basin is large enough to handle peak flow rates without causing surface flooding.
- **Sediment and Debris Removal**: Catch basins are equipped with sumps, or pits, at the bottom to trap sediment. Regular maintenance and cleaning are required to prevent clogging and ensure effective operation.
- **Outlet Design**: The outlet pipe size must be carefully selected to balance the removal of stormwater with the prevention of sediment escape. Grates or covers are also designed to prevent large debris from entering the basin while allowing water to flow through.

Culverts serve as conduits for stormwater to flow under roadways, railways, or other barriers. They are typically made from materials such as concrete, metal, or plastic. The design and installation of culverts are critical for maintaining natural water flow, preventing erosion, and ensuring road safety. Key design considerations for culverts include:

- **Hydraulic Capacity**: The size and shape of the culvert are determined based on hydraulic analysis to ensure it can convey stormwater efficiently without causing upstream flooding or excessive velocity that could lead to downstream erosion.
- **Material Selection**: The choice of material depends on factors such as the expected flow, soil characteristics, environmental conditions, and load-bearing requirements. Durability and maintenance needs are also important considerations.
- **Inlet and Outlet Control**: The design of the culvert's inlet and outlet affects its hydraulic performance and the potential for blockage. Energy dissipators or stilling basins may be required at the outlet to reduce flow velocity and prevent erosion.
- **Fish Passage**: In areas where culverts intersect natural streams, design modifications such as embedded culverts or fish ladders may be necessary to facilitate the passage of aquatic organisms and maintain ecosystem connectivity.

The integration of catch basins and culverts into the stormwater collection network requires careful planning and coordination with the overall stormwater management strategy. This includes considerations for future development, climate change impacts, and regulatory compliance. The design process involves detailed hydrologic and hydraulic modeling to simulate storm events, assess system performance, and identify optimal locations for these components. By effectively capturing, conveying, and treating stormwater, catch basins and culverts play a vital role in minimizing flood risk, protecting infrastructure, and enhancing water quality in urban environments.

Wastewater Systems Design and Management

In the realm of wastewater systems, the design of sewer networks plays a pivotal role in ensuring the efficient collection and conveyance of wastewater to treatment facilities. This process is critical for maintaining public health, protecting the environment, and supporting urban development. The design of sewer systems requires a comprehensive understanding of hydraulic engineering principles, environmental considerations, and regulatory standards. One of the primary considerations in sewer design is the determination of flow rates, which are influenced by a variety of factors including population density, water usage patterns, and the area's topography.

Flow estimation in sewer system design is a complex process that involves calculating the average and peak wastewater flows from residential, commercial, industrial, and institutional sources. The estimation of these flows is crucial for selecting the appropriate size of sewer pipes and other infrastructure components to prevent overflows and ensure efficient operation under varying conditions. The design flow rates are typically determined using empirical formulas that take into account the contributing population, per capita water usage, and infiltration/inflow of extraneous water into the system. For instance, the design of sanitary sewers often employs the formula $Q = P \times q \times (1 + i)$, where Q is the design flow rate, P is the population served, q is the average per capita wastewater contribution, and i is a factor accounting for infiltration and inflow.

In addition to flow estimation, the management of combined sewer systems presents unique challenges. Combined sewers, which collect both stormwater runoff and sanitary wastewater in a single pipe system, are particularly common in older urban areas. These systems are designed to convey the mixture to a wastewater treatment plant under normal conditions. However, during heavy rainfall events, the volume of water can exceed the system's capacity, leading to combined sewer overflows (CSOs) that discharge untreated wastewater directly into rivers, lakes, or oceans. Managing CSOs effectively requires a strategic approach that may include the implementation of green infrastructure to reduce stormwater runoff, the construction of overflow storage facilities, and the separation of sanitary and storm sewers where feasible.

The design and management of wastewater systems also necessitate a thorough understanding of the regulatory framework governing water quality and environmental protection. Engineers must ensure that sewer designs comply with national and local standards, which may dictate specific requirements for flow rates, overflow control measures, and effluent quality. Additionally, the integration of sustainable practices into sewer system design and management is increasingly important as communities seek to mitigate environmental impacts, conserve water resources, and adapt to changing climate conditions.

As we delve further into the intricacies of wastewater systems, it becomes evident that the engineering decisions made in the design and management phases have far-reaching implications for public health, environmental sustainability, and urban resilience. The complexity of these systems requires a multidisciplinary approach, combining expertise in hydraulic engineering, environmental science, and urban planning to develop solutions that are efficient, sustainable, and compliant with regulatory standards.

The meticulous planning and execution of sewer system designs are further complicated by the need to accommodate future growth and development within urban areas. This foresight involves not only sizing the infrastructure to handle anticipated increases in wastewater volumes but also ensuring that the network's layout is flexible enough to integrate with future expansions or modifications. Such planning requires an in-depth analysis of urban development trends, demographic shifts, and potential changes in water usage patterns, necessitating a dynamic approach to sewer system design that can adapt to evolving needs.

Moreover, the adoption of advanced technologies and modeling tools has become integral to the design and management of wastewater systems. Hydraulic modeling software, for instance, allows engineers to simulate the flow of wastewater through the network under various scenarios, including peak flow events and potential system failures. These models are invaluable for identifying bottlenecks, assessing the impact of infrastructure changes, and planning for capacity upgrades. Additionally, Geographic Information Systems (GIS) are employed to map the sewer network, facilitating the visualization of the system's layout, identifying critical assets, and supporting maintenance and operational decisions.

The environmental impact of wastewater systems is also a critical consideration, with engineers increasingly turning to green infrastructure solutions to enhance system performance while minimizing ecological footprints. Techniques such as permeable pavements, green roofs, and rain gardens not only reduce stormwater runoff and alleviate pressure on combined sewer systems but also contribute to urban green spaces, improving air quality and enhancing the urban landscape. These solutions exemplify the shift towards more sustainable urban water management practices, emphasizing the role of wastewater systems in achieving broader environmental and societal goals.

Effective management of wastewater systems extends beyond the technical and environmental aspects, encompassing financial and social considerations as well. The economic viability of sewer system projects, including construction, operation, and maintenance costs, must be carefully evaluated to ensure that they deliver value for money and do not impose undue financial burdens on communities. Public engagement and education are equally important, as community support and understanding can significantly influence the success of wastewater management initiatives. By fostering a sense of shared responsibility and encouraging water-conserving behaviors, engineers and city planners can enhance the efficiency and sustainability of wastewater systems.

The design and management of wastewater systems represent a complex interplay of engineering, environmental, economic, and social factors. The challenges inherent in these systems demand a comprehensive and forward-looking approach, leveraging the latest technologies and embracing sustainable practices to meet the needs of both current and future generations. As urban areas continue to grow and evolve, the role of engineers in developing innovative, resilient, and sustainable wastewater management solutions becomes increasingly critical, underscoring the importance of continuous learning, adaptation, and collaboration in the field of water resources and environmental engineering.

Groundwater

Flow Principles and Groundwater Modeling

Darcy's Law is a fundamental principle that describes the flow of fluid through a porous medium. The law states that the flow rate through a porous medium is proportional to the hydraulic gradient, assuming the flow is laminar and the medium is homogeneous and isotropic. Mathematically, Darcy's Law can be expressed as:

$$Q = -KA\frac{\Delta h}{L}$$

where Q is the volumetric flow rate (L3/T), K is the hydraulic conductivity of the medium (L/T), A is the cross-sectional area perpendicular to the flow direction (L2), Δh is the difference in hydraulic head over the length L of the medium (L), and the negative sign indicates that flow occurs from higher to lower hydraulic head.

Hydraulic conductivity (K) is a measure of a material's ability to transmit water. It depends on both the intrinsic permeability of the material and the properties of the fluid, primarily its viscosity and density. Hydraulic conductivity is a critical parameter in groundwater studies as it influences how quickly groundwater can move through the subsurface environment. Values of K vary widely among different types of geological materials, from highly permeable gravel layers that can transmit water rapidly to low-permeability clay layers that impede water movement.

Groundwater flow modeling is a computational tool used to simulate and predict the flow of groundwater through aquifers. These models are essential for managing water resources, predicting the spread of contaminants, and understanding natural processes such as recharge and discharge in aquifer systems. Groundwater models typically solve the groundwater flow equation, which is based on Darcy's Law and the principle of mass conservation. The flow equation for a homogeneous, isotropic aquifer in three dimensions is given by:

$$\frac{\partial}{\partial x}\left(K\frac{\partial h}{\partial x}\right) + \frac{\partial}{\partial y}\left(K\frac{\partial h}{\partial y}\right) + \frac{\partial}{\partial z}\left(K\frac{\partial h}{\partial z}\right) = S_s\frac{\partial h}{\partial t}$$

where h is the hydraulic head (L), S_s is the specific storage of the aquifer material (1/L), and t is time (T). This equation is solved under appropriate initial and boundary conditions to predict changes in hydraulic head over time and space.

Modeling groundwater flow requires detailed information about the aquifer geometry, hydraulic properties of the aquifer materials (such as hydraulic conductivity and specific storage), and boundary conditions that represent interactions with rivers, lakes, wells, or other features. Groundwater models can range from simple analytical solutions for idealized conditions to complex numerical models that discretize the aquifer into a grid or mesh and solve the flow equation for each cell using computational methods.

The accuracy of groundwater flow models depends on the quality of the input data, the appropriateness of the model assumptions (e.g., homogeneity, isotropy), and the numerical methods used to solve the flow equations. Calibration, a process of adjusting model parameters to match observed data, and validation, a process of comparing model predictions with independent data sets, are critical steps in ensuring that a groundwater model is reliable and useful for decision-making.

Groundwater flow modeling serves as a powerful tool in the hands of engineers and hydrologists, enabling them to assess water resource availability, design remediation strategies for contaminated sites, and evaluate the impacts of groundwater extraction or recharge projects. Through the application of Darcy's Law and advanced computational techniques, these professionals can predict groundwater behavior under various scenarios, supporting sustainable management of this vital resource.

Wells and Drawdown Analysis

In the context of groundwater management and engineering, **well design** and **drawdown analysis** are critical for ensuring sustainable water extraction and minimizing environmental impact. Well design involves several considerations, including determining the optimal diameter, depth, and screen placement to access groundwater efficiently while maintaining the aquifer's integrity. The design process must account for the aquifer's hydraulic properties, such as transmissivity and storativity, to predict how the well will perform under various pumping rates.

The **pumping test**, also known as an aquifer test, plays a pivotal role in understanding these hydraulic properties. During a pumping test, water is extracted from a well at a constant rate, and the response of the groundwater system is observed in terms of water level changes both in the pumped well and in observation wells. The data collected from these tests are analyzed to estimate the aquifer's hydraulic conductivity (K) and specific yield (S_y) or storativity (S), which are essential for designing a sustainable pumping regime and predicting long-term impacts on the aquifer.

Drawdown refers to the lowering of the groundwater level in a well as a result of pumping. The drawdown in a well and its surrounding observation wells provides insight into the aquifer's behavior and its ability to replenish. The relationship between the pumping rate (Q) and the drawdown (s) can be described by the well-known Thiem equation for steady-state flow to a well in a confined aquifer:

$$Q = 2\pi T \frac{h_0 - h}{\ln(r/r_w)}$$

where T is the transmissivity of the aquifer, h_0 is the original hydraulic head, h is the hydraulic head at distance r from the well, and r_w is the radius of the well. This equation helps in understanding the impact of pumping on groundwater levels and is crucial for designing wells that do not overexploit the aquifer.

Drawdown analysis involves using the data from pumping tests to model the aquifer's response to different pumping scenarios. This can include predictive modeling for various pumping rates, durations, and well configurations to ensure that the groundwater extraction does not lead to undesirable consequences such as excessive drawdown, aquifer depletion, or land subsidence. Advanced models may also consider the effects of boundary conditions, recharge rates, and aquifer heterogeneity.

For sustainable well design and operation, engineers must also consider the **specific capacity** of a well, which is the rate of water flow per unit drawdown (Q/s). This parameter is a measure of well efficiency and is influenced by both the well design and the aquifer properties. A well with a high specific capacity requires less energy for pumping, making it more cost-effective and environmentally sustainable in the long run.

Well design and drawdown analysis are integral components of groundwater engineering, requiring a thorough understanding of aquifer hydraulics, well hydraulics, and environmental sustainability principles. By carefully designing wells and analyzing drawdown, engineers can ensure that groundwater resources are utilized efficiently and sustainably, supporting both current and future water needs without compromising the health of the aquifer system.

Water Quality

Surface and Groundwater Quality Parameters

Evaluating the **quality of surface and groundwater** is essential for ensuring the safety and sustainability of water resources. This evaluation involves analyzing physical, chemical, and biological parameters, each of which provides insights into the water's characteristics and potential impacts on human health and the environment.

Physical parameters include temperature, turbidity, color, and total suspended solids (TSS). Temperature is a critical factor affecting water's chemical and biological processes, with warmer temperatures accelerating many reactions. Turbidity, the measure of water's clarity, can indicate the presence of suspended particles that may harbor pathogens or interfere with light penetration, affecting aquatic ecosystems. Color, often caused by dissolved organic materials, can indicate contamination sources, while TSS measures the concentration of suspended particles, affecting aquatic life and water filtration processes.

Chemical parameters encompass a wide range of substances, including pH, dissolved oxygen (DO), hardness, metals, and nutrients like nitrogen and phosphorus. The pH level indicates water's acidity or alkalinity, influencing metal solubility and organism survival. Dissolved oxygen is crucial for aquatic life; its concentration can reflect water quality and pollution levels. Hardness, primarily caused by calcium and magnesium ions, affects water's suitability for industrial and domestic use. Metals such as lead, arsenic, and mercury are toxic contaminants that can enter waterways through natural processes and human activities. Nutrients, while essential for aquatic ecosystems, can lead to eutrophication and harmful algal blooms if concentrations become excessive.

Biological parameters focus on the presence and abundance of microorganisms, including bacteria, viruses, protozoa, and algae. Indicator organisms, such as E. coli, signal fecal contamination and potential pathogen presence. The diversity and population of aquatic organisms can also indicate water quality, with certain species being more tolerant of pollution than others.

To assess these parameters, engineers and environmental scientists employ various testing methods, including spectrophotometry for chemical analysis, microscopy for identifying microorganisms, and physical sensors for measuring temperature and turbidity. These assessments help in determining water treatment needs, assessing compliance with environmental regulations, and monitoring the health of aquatic ecosystems.

For instance, the **analysis of dissolved oxygen** utilizes the Winkler method for titrimetric determination, essential for evaluating water's ability to support aerobic life. The **measurement of pH** is conducted using pH meters, providing immediate data critical for managing water chemistry. **Turbidity** is measured using nephelometers, with results guiding treatment processes to ensure water clarity.

Understanding and managing the quality of surface and groundwater require a comprehensive approach, integrating physical, chemical, and biological assessments to protect human health and the environment. Through diligent monitoring and analysis, engineers can develop effective strategies for water treatment, pollution control, and resource management, ensuring the sustainability of vital water resources.

Basic Chemistry in Water Quality Assessment

Understanding the basic chemistry of water involves delving into the concepts of **pH**, **alkalinity**, and **hardness**, each playing a pivotal role in water quality assessment. These parameters are crucial for engineers and environmental scientists to evaluate the suitability of water for various uses, including drinking, industrial processes, and aquatic life support.

pH is a measure of the hydrogen ion concentration in water, representing its acidity or alkalinity on a scale ranging from 0 to 14. A pH of 7 is neutral, values less than 7 indicate acidity, and values greater than 7 denote alkalinity. The pH level of water can significantly affect the solubility of minerals and metals, as well as the effectiveness of disinfection processes. For instance, high pH levels can lead to the precipitation of calcium and magnesium, while low pH levels can increase the solubility of toxic metals like lead and copper, making them more available and potentially harmful to consumers. The pH of water is determined using the equation:

$$\mathrm{pH} = -\log_{10}[H^+]$$

where $[H^+]$ is the concentration of hydrogen ions in moles per liter.

Alkalinity, on the other hand, measures the water's capacity to neutralize acids, which is essential for maintaining stable pH levels, especially in natural water bodies. It is primarily contributed by bicarbonates (HCO_3^-), carbonates (CO_3^{2-}), and, in some cases, hydroxides (OH^-). Alkalinity is critical for aquatic life, as sudden changes in pH can be detrimental. It acts as a buffer, protecting the water body from pH swings that could result from acid rain or industrial discharges. The total alkalinity of water is quantified by titrating with a standard acid solution to a designated pH endpoint, typically using phenolphthalein and methyl orange indicators to represent different alkalinity fractions.

Hardness is primarily a measure of the concentration of divalent metal ions, such as calcium (Ca^{2+}) and magnesium (Mg^{2+}), in water. These minerals originate from the dissolution of limestone and other minerals in the water's source. Hard water can lead to scale formation in boilers, water heaters, and pipes, reducing their efficiency and lifespan. Conversely, very soft water can lead to increased wear on metallic surfaces due to its higher corrosivity. The hardness of water is often expressed in terms of the equivalent concentration of calcium carbonate $(CaCO_3)$ in milligrams per liter (mg/L) and is determined using complexometric titration with ethylenediaminetetraacetic acid (EDTA) or by using atomic absorption or emission spectroscopy.

The interplay between pH, alkalinity, and hardness is fundamental in water treatment processes, environmental monitoring, and ensuring the safety and functionality of water infrastructure. For engineers and environmental scientists, a thorough understanding of these parameters, along with their measurement and interpretation, is essential for designing and managing water treatment systems, assessing the impact of discharges on natural water bodies, and complying with environmental and public health regulations. Through meticulous monitoring and management of water chemistry, professionals can ensure the provision of safe, high-quality water for all intended uses, while also protecting the natural environment.

Testing and Standards

Water and Wastewater Testing

Continuing from the foundational understanding of water quality parameters, the focus shifts to the critical tests employed in water and wastewater analysis, which are pivotal for ensuring compliance with environmental standards and safeguarding public health. Among these, the Biological Oxygen Demand (BOD), Total Suspended Solids (TSS), and water potability tests stand out for their widespread application and significance in assessing water quality.

Biological Oxygen Demand (BOD) is a measure of the quantity of oxygen consumed by microorganisms during the decomposition of organic matter in water. It is an essential indicator of the organic pollution level in water and wastewater. The BOD test involves incubating a sealed sample of water at a controlled temperature (20°C) for a specific period, typically five days (BOD₅). The difference in dissolved oxygen (DO) concentration between the initial and final state of the sample is measured to determine the BOD value. The equation for BOD can be represented as:

$$BOD = (D_1 - D_2) - (B_1 - B_2)$$

where D_1 and D_2 are the DO concentrations of the diluted sample at the start and end of the incubation period, respectively, and B_1 and B_2 are the DO concentrations of the blank (deionized water) at the start and end of the incubation period. This test is critical for wastewater treatment facilities to ensure that the effluent discharged into natural water bodies does not detrimentally affect aquatic life or contribute to eutrophication.

Total Suspended Solids (TSS) quantifies the particles suspended in water or wastewater, which can include a wide range of materials such as silt, decaying plant matter, industrial wastes, and sewage. High levels of TSS can harm aquatic life, interfere with the penetration of sunlight, and contribute to the spread of pathogens. The TSS is determined by filtering a known volume of water through a pre-weighed filter, drying the filter in an oven at 103-105°C, and then reweighing it. The increase in the filter's weight corresponds to the mass of the suspended solids in the water sample. The TSS concentration is calculated using the formula:

$$TSS(mg/L) = \frac{(W_2 - W_1) \times 1000}{V}$$

where W_1 is the weight of the filter before filtration, W_2 is the weight of the filter after drying, and V is the volume of the water sample in liters.

Water potability analysis encompasses a series of tests to assess the suitability of water for human consumption. These tests evaluate various parameters, including microbial contaminants (e.g., E. coli, Enterococci), chemical pollutants (e.g., nitrates, heavy metals), and physical characteristics (e.g., pH, turbidity). The Safe Drinking Water Act (SDWA) in the United States sets the maximum contaminant levels (MCLs) for these substances to ensure water safety. For instance, the presence of E. coli is a direct indicator of fecal contamination, and its detection in a water supply necessitates immediate remedial action. Chemical analyses for potability often involve spectrophotometry, chromatography, and mass spectrometry to identify and quantify pollutants with precision.

The integration of BOD, TSS, and water potability tests into regular monitoring routines is indispensable for water resource management. These tests provide actionable data that guide the treatment processes, ensuring that both treated wastewater and drinking water meet the requisite quality standards. Moreover, they serve as benchmarks for environmental compliance, helping to prevent pollution and protect ecosystems. For engineers and environmental scientists, proficiency in conducting these tests and interpreting their results is crucial for designing effective water treatment solutions and safeguarding public health. Through meticulous application of these testing protocols, professionals can address the challenges of water pollution and scarcity, contributing to the sustainable management of this vital resource.

Air and Noise Standards

Transitioning from water quality assessment to the domain of air and noise standards, it's imperative to understand the regulatory frameworks and testing protocols that govern these environmental factors. Air quality and noise pollution are critical aspects of environmental engineering, with direct impacts on public health, ecosystem balance, and compliance with legal standards. The methodologies for monitoring, assessing, and mitigating air and noise pollution are underpinned by a complex array of standards and regulations designed to protect and preserve environmental integrity.

Air quality standards are primarily focused on controlling the concentration of pollutants in the atmosphere, which include, but are not limited to, particulate matter (PM10 and PM2.5), nitrogen dioxide (NO_2), sulfur dioxide (SO_2), carbon monoxide (CO), ozone (O_3), and lead (Pb). The

Environmental Protection Agency (EPA) in the United States sets National Ambient Air Quality Standards (NAAQS) for these pollutants to safeguard public health and the environment. The assessment of air quality involves the collection of air samples, monitoring pollutant levels using both stationary and mobile stations, and analyzing these samples with techniques such as gas chromatography, mass spectrometry, and spectrophotometry to quantify the concentrations of various pollutants.

For particulate matter, for instance, the gravimetric method is a standard protocol, where air is drawn through a filter, and the mass of the particles is measured. The concentration of PM is then calculated based on the volume of air sampled. For gases like ozone, nitrogen dioxide, and sulfur dioxide, chemiluminescence and UV fluorescence are commonly employed methods, providing high sensitivity and specificity for detecting these pollutants at low concentrations. The EPA's Air Quality Index (AQI) is a tool used to communicate how polluted the air currently is or how polluted it is forecast to become, with different colors indicating the health implications of the air quality levels.

Noise pollution, on the other hand, is assessed through sound level measurements, typically using sound level meters that measure the intensity of sound in decibels (dB). These devices are equipped to measure sound pressure levels over a range of frequencies, accurately capturing the ambient noise environment. The standards for noise pollution are set to protect against hearing loss, sleep disturbance, and other health effects. Regulations may specify maximum allowable noise levels for different environments, such as residential areas, commercial zones, and industrial sites, both during the day and at night. The Occupational Safety and Health Administration (OSHA) sets legal limits on noise exposure in the workplace, with the permissible exposure limit set at 90 dBA for an 8-hour shift, and the implementation of a hearing conservation program when worker exposure is at or above 85 dBA for an 8-hour shift.

The assessment of noise pollution involves not only measuring the sound levels but also analyzing the frequency spectrum to identify dominant frequencies and their sources. This analysis aids in the development of mitigation strategies, such as soundproofing, the use of noise barriers, and the implementation of quieter technologies. In addition to direct measurements, modeling and simulation tools are also utilized to predict noise levels from proposed developments or changes in land use, facilitating proactive management of noise pollution.

Compliance with air and noise standards is enforced through a combination of regulatory oversight, permitting processes for industrial and construction activities, and public education initiatives aimed at reducing pollution at the source. Continuous monitoring and reporting are essential components of compliance, with data being used to inform policy decisions, guide the allocation of resources for pollution control, and evaluate the effectiveness of mitigation measures.

In the context of preparing for the FE Civil Exam, understanding the principles of air and noise pollution assessment, the regulatory landscape, and the technical methodologies for measuring and analyzing pollutants is crucial. This knowledge not only contributes to a well-rounded environmental engineering education but also equips future engineers with the competencies needed to address some of the most pressing environmental challenges of our time.

Water and Wastewater Treatment

Drinking Water Treatment Methods

In the realm of drinking water treatment, coagulation stands as a pivotal initial step, designed to enhance the efficiency of subsequent filtration processes. This method involves the addition of chemicals, known as coagulants, which possess a positive charge. The primary objective of these coagulants is to neutralize the negative charge of particles suspended in water, such as dirt, organic matter, and bacteria, thereby mitigating their stability. Once these particles are destabilized, they aggregate to form larger particles, termed "flocs," through a process known as flocculation. The flocs are more readily removed from the water during the filtration stage. Common coagulants include aluminum sulfate (alum) and ferric chloride, each chosen based on the specific characteristics of the water being treated and the desired outcome of the coagulation process.

Following coagulation, water undergoes filtration, a critical phase aimed at removing particles from water, including those formed during coagulation. Filtration can be achieved through various means, including rapid sand filters, slow sand filters, and membrane filters, each offering distinct advantages and suitability for different water qualities and treatment goals. Rapid sand filters, for instance, are characterized by their ability to filter water at a faster rate, making them suitable for large-scale water treatment facilities. They operate by allowing water to pass through layers of sand and gravel, trapping and removing particles. Conversely, slow sand filters, though operating at a slower filtration rate, are lauded for their biological layer, known as a "schmutzdecke," which forms on the surface of the filter and aids in the removal of additional contaminants through biological processes.

Softening, another crucial aspect of water treatment, specifically targets the reduction of water hardness, which is predominantly caused by the presence of calcium and magnesium ions. Hard water can lead to various issues, including scaling in pipes and appliances, reduced effectiveness of soaps and detergents, and challenges in water heating systems. The softening process often involves the exchange of hard ions for softer ones, such as sodium or potassium, through ion exchange resins or the addition of chemicals that precipitate hard ions out of solution. This not only prevents scaling but also enhances the aesthetic quality of water, making it more palatable and suitable for everyday use.

Disinfection, a vital final step in the drinking water treatment process, ensures the elimination of pathogenic microorganisms that pose health risks to consumers. This is typically achieved through the application of chemical disinfectants, such as chlorine, chloramines, or ozone, which effectively kill or deactivate harmful bacteria, viruses, and protozoa. Alternatively, physical methods like ultraviolet (UV) light exposure can also be employed to achieve disinfection, offering the advantage of leaving no residual chemicals in the treated water. The choice of disinfection method depends on various factors, including the quality of the source water, the presence of residual contaminants, and the specific health and safety standards that the treated water must meet.

The effectiveness of the disinfection process is critically dependent on the preceding treatment stages. Proper coagulation and filtration significantly reduce the load of organic matter and microorganisms, thereby enhancing the efficiency of disinfectants and minimizing the formation of disinfection by-products (DBPs). DBPs, such as trihalomethanes (THMs) and haloacetic acids (HAAs), are formed when disinfectants react with residual organic matter in the water. These compounds are of concern due to their potential health risks, including carcinogenic effects. Therefore, optimizing the coagulation, filtration, and softening processes is essential to ensure that disinfection can be conducted effectively while minimizing the formation of harmful by-products.

In addition to chemical and physical methods of disinfection, advanced oxidation processes (AOPs) have emerged as a powerful tool for water treatment. AOPs involve the generation of highly reactive species, such as hydroxyl radicals, which can effectively degrade a wide range of organic pollutants and microorganisms. These processes are particularly useful for treating water with persistent organic pollutants that are resistant to conventional treatment methods. AOPs can be integrated into the treatment train as a complementary step to conventional disinfection, offering an additional barrier against contaminants.

Monitoring and control are integral components of the drinking water treatment process. Continuous monitoring of water quality parameters, such as turbidity, pH, and disinfectant levels, is crucial to ensure that the treatment objectives are being met and that the water is safe for consumption. Automated control systems can adjust the dosages of coagulants and disinfectants in real-time based on water quality data, optimizing the treatment process and ensuring consistent compliance with regulatory standards.

The selection of treatment methods and the design of the treatment process must consider the specific characteristics of the source water, the regulatory requirements, and the goals for water quality. Each water treatment facility is unique, and the treatment process may involve a combination of different technologies and methods tailored to achieve the desired water quality outcomes. The ultimate goal is to provide safe, reliable, and aesthetically pleasing drinking water that meets the needs and expectations of the community it serves.

As the field of water treatment continues to evolve, new technologies and approaches are being developed to address emerging contaminants and to improve the efficiency and sustainability of water treatment processes. Engineers and water quality professionals play a critical role in integrating these advancements into practice, ensuring that drinking water remains safe and accessible for all.

Wastewater Treatment Processes

Wastewater treatment is a critical component of environmental engineering, focusing on the removal of contaminants from wastewater to produce an effluent that is safe for discharge into the environment or for reuse. Among the various processes involved, biological treatment methods play a pivotal role, particularly through the use of activated sludge systems. These systems leverage microbial communities to decompose organic matter and other pollutants, transforming wastewater into a less harmful state.

The activated sludge process is a type of secondary treatment that involves aerating wastewater to encourage the growth and activity of bacteria and other microorganisms. These microorganisms form a flocculent sludge that absorbs organic pollutants, effectively reducing the biochemical oxygen demand (BOD) and total suspended solids (TSS) in the wastewater. The process begins with the aeration tank, where air or oxygen is injected into the mixed liquor—comprising wastewater and recycled sludge—to provide the necessary oxygen for aerobic microorganisms to thrive and consume the organic matter.

Key to the effectiveness of the activated sludge process is maintaining the optimal conditions for microbial growth, including temperature, pH, and oxygen levels. The mixed liquor suspended solids (MLSS) concentration is another critical parameter, indicating the biomass concentration in the aeration tank. Control of these conditions ensures the efficient breakdown of organic pollutants, while preventing issues such as bulking or foaming, which can disrupt the treatment process.

Following aeration, the mixed liquor flows into a secondary clarifier, where the activated sludge is allowed to settle, separating the treated effluent from the biomass. The clarified effluent can then undergo further treatment or be discharged, while a portion of the settled sludge is recycled back to the aeration tank to maintain the active biomass. The excess sludge, or waste activated sludge (WAS), is removed from the process and subjected to sludge treatment and disposal methods, which may include digestion, dewatering, and land application or incineration.

Sludge management is an integral part of the activated sludge process, addressing the handling, treatment, and disposal of the sludge produced. Effective sludge management is essential not only for the sustainability of the wastewater treatment process but also for minimizing environmental impacts. Sludge treatment options aim to reduce volume, stabilize organic matter, and eliminate pathogens to produce a safe and manageable waste product. Anaerobic digestion is a common sludge treatment method, converting organic matter into biogas, which can be used as a renewable energy source, and a stabilized sludge that is easier to dewater and dispose of.

The activated sludge process and subsequent sludge management techniques exemplify the complex interplay of biological, chemical, and physical principles underlying wastewater treatment. Through the optimization of these processes, engineers can effectively reduce the environmental impact of wastewater, contributing to the protection of water resources and public health.

Anaerobic digestion, while effective, represents just one facet of the multifaceted approach required for comprehensive sludge management. Following digestion, the dewatering process further reduces the sludge volume, making it more suitable for disposal or use as a soil amendment. This step often involves mechanical equipment such as belt filter presses, centrifuges, or screw presses, which exert pressure on the digested sludge to expel water. The choice of dewatering technology depends on the sludge characteristics, the desired dryness level, and economic considerations, including both operational and maintenance costs.

Moreover, advanced thermal processes, such as incineration or pyrolysis, offer pathways for sludge volume reduction and the recovery of energy. Incineration oxidizes organic materials in the sludge, reducing its mass and volume significantly and generating heat that can be converted into electricity. Pyrolysis, on the other hand, decomposes organic sludge components at high temperatures in the absence of oxygen, producing biochar, oils, and syngas. These products can be utilized as renewable energy sources or as materials in various industrial applications, contributing to the circular economy.

The management of sludge also encompasses the careful consideration of environmental regulations and standards, which dictate the permissible methods for sludge disposal and reuse. Land application of treated sludge, for instance, is subject to stringent criteria to prevent soil and water contamination. This method recycles organic matter and nutrients back into the soil, enhancing its fertility and structure. However, it requires thorough testing and treatment of sludge to eliminate pathogens and minimize the presence of heavy metals and other hazardous substances.

The integration of sludge treatment and disposal into the overall design of wastewater treatment plants necessitates a holistic approach, considering not only the technical and economic aspects but also the environmental and social implications. Engineers must evaluate the life cycle impacts of sludge management options, including greenhouse gas emissions, energy consumption, and potential benefits such as energy recovery and resource recycling. This evaluation helps in selecting the most sustainable and cost-effective sludge management strategy, aligned with the principles of environmental stewardship and regulatory compliance.

In conclusion, the activated sludge process, complemented by advanced sludge management techniques, embodies the intricate balance between engineering innovation and environmental responsibility. By optimizing these processes, engineers play a pivotal role in enhancing the sustainability of wastewater treatment systems, safeguarding water quality, and contributing to the broader goals of public health protection and environmental conservation. Through continuous improvement and adoption of best practices, the wastewater treatment sector can address the challenges of sludge management, turning waste into valuable resources and achieving greater efficiency in the protection of our water resources.

Chapter 11: Structural Engineering

Statically Determinant Structures Analysis

Structural Analysis Basics

In the realm of structural engineering, understanding the principles of shear, bending moment, and axial force calculations is paramount for the analysis of statically determinant structures. These fundamental concepts form the backbone of structural analysis, enabling engineers to predict the behavior of structures under various loads and ultimately ensuring their safety and stability.

Shear forces in a beam or structural element are a result of external loads, such as live loads and dead loads, applied transversely to the axis of the element. These forces cause the material to slide over itself along the plane of the force, which can significantly affect the integrity of the structure. The calculation of shear forces is essential for the design of beams and for ensuring that the shear capacity of the material is not exceeded. The shear force at any given point along a beam is determined by summing the vertical forces on one side of the point. This is mathematically represented as $V = \sum F_{vertical}$, where V is the shear force and $F_{vertical}$ represents the vertical forces.

Bending moments, on the other hand, are moments that cause the structure to bend. They are generated by external loads, moments, or reactions which act at a distance from the cross section under consideration. The bending moment at any point along a beam is calculated by taking the sum of the moments about that point. The moment is positive if it causes compression at the top of the beam, which is the convention for most engineering applications. The bending moment M at a point can be calculated using the equation $M = \sum (F \times d)$, where F is the force applied and d is the distance from the point to the force's line of action.

Axial forces are forces that are applied along the longitudinal axis of a structural element, causing it to either stretch (tension) or shorten (compression). These forces are critical in the analysis of columns, struts, and ties, as they directly influence the element's ability to support loads without buckling or failing. The axial force N in a member can be determined by summing the longitudinal forces acting on the member, represented by the equation $N = \sum F_{\text{longitudinal}}$.

The interplay between shear forces, bending moments, and axial forces is a key aspect of structural analysis. For instance, in a simply supported beam subjected to a uniform distributed load, the shear force varies linearly along the length of the beam, and the bending moment follows a quadratic distribution, reaching its maximum at the midpoint of the beam. The analysis of these forces and moments involves creating shear and moment diagrams, which graphically represent how shear forces and bending moments vary along the length of a beam. These diagrams are invaluable tools for identifying points of maximum stress and for designing structural elements that are both efficient and safe.

The equilibrium equations, derived from Newton's First Law, are fundamental in determining the internal forces and moments in a structure. For a structure to be in equilibrium, the sum of the horizontal forces, the sum of the vertical forces, and the sum of the moments about any point must each be equal to zero. These conditions can be expressed as $\sum F_x = 0$, $\sum F_y = 0$, and $\sum M = 0$, respectively. Applying these equilibrium conditions to a free-body diagram of the structure allows for the calculation of unknown forces and moments, facilitating the comprehensive analysis of statically determinant structures.

In statically determinant structures, the application of these principles enables the complete determination of internal forces and moments from the external loads and support conditions without the need for additional equations of compatibility. This contrasts with statically indeterminate structures, where the internal forces and moments cannot be found using the equations of equilibrium alone, and additional compatibility equations, based on the deformation of the structure, are required.

The accurate calculation of shear forces, bending moments, and axial forces is crucial for the design and analysis of structural elements. It ensures that structures can withstand the applied loads without experiencing failure or excessive deformation, thereby safeguarding the well-being of the public and the longevity of the built environment. Through meticulous analysis and adherence to established engineering principles, structural engineers can design structures that not only meet the required safety standards but also optimize material use and cost-efficiency, reflecting the profound impact of these fundamental concepts on the field of structural engineering.

Practical Applications: Analyzing Trusses, Beams, Frames

Analyzing trusses, beams, and frames under various loading and support conditions requires a comprehensive understanding of structural behavior and the application of static equilibrium principles. Each structural element and system has unique characteristics that influence how forces and moments are distributed and resisted.

Trusses are commonly used in bridges, roofs, and towers. They consist of members arranged in connected triangles. When analyzing trusses, the method of joints and the method of sections are two fundamental approaches. The **method of joints** involves isolating each joint and applying the equilibrium equations $\sum F_x = 0$ and $\sum F_y = 0$ to solve for the unknown forces in the members. This method is particularly effective for trusses with a simple configuration or when the forces in all the members connected to a specific joint are required. On the other hand, the **method of sections** involves cutting through the truss to expose a section containing up to three unknown forces and then applying the equilibrium equations, including the moment equilibrium equation $\sum M = 0$, to solve for these forces. This approach is efficient for determining forces in specific members without needing to analyze the entire truss.

Beams are structural elements that resist loads primarily through bending. The analysis of beams involves determining the shear force V and bending moment M at any point along the beam's length. The relationship between the load on the beam $w(x)$, the shear force, and the bending moment is given by the differential relationships $dV/dx = -w(x)$ and $dM/dx = V$. Integrating these expressions allows for the construction of shear and moment diagrams, which graphically represent the variation of shear force and bending moment along the length of the beam. These diagrams are crucial for identifying the maximum bending moment and shear force values, which directly influence the design and selection of beam dimensions and reinforcement.

Frames are structures consisting of beams and columns connected to form a rigid structure. Unlike trusses, frames experience bending moments and shear forces at the joints and along the members. The analysis of frames often requires considering the additional moments generated at the connections due to the rigidity of the joints. Simplified frame analysis can be performed using the assumption of pinned connections to reduce complexity, but more accurate analysis typically involves the use of moment distribution methods or software-based finite element analysis to account for the stiffness of the connections and the distribution of moments and shears throughout the frame.

When analyzing structures under various loading and support conditions, it is essential to consider the effects of different types of loads, such as dead loads, live loads, wind loads, and seismic loads. Each type of load has unique characteristics and may produce different responses in the structure. For example, live loads are variable and can change the distribution of forces within the structure, while wind and seismic loads can introduce lateral forces that challenge the structure's stability.

The support conditions of a structure significantly influence its ability to resist loads. Fixed supports restrain both translation and rotation, providing greater stability but also introducing bending moments at the supports. Pinned supports allow rotation, which can reduce bending moments but may require additional bracing to maintain structural stability. Roller supports offer a single point of resistance, allowing both rotation and horizontal movement, which can be useful in accommodating thermal expansion or ground settlement.

In conclusion, the analysis of trusses, beams, and frames under various loading and support conditions is a fundamental aspect of structural engineering. By applying static equilibrium principles and considering the unique characteristics of each structural element and system, engineers can design structures that safely and efficiently resist the imposed loads, ensuring their stability and longevity.

Deflection of Statically Determinant Structures

Deflection Methods for Beams and Trusses

The moment-area method, another pivotal technique for analyzing deflection in beams and trusses, leverages the relationship between moments and the curvature of the beam. According to the theorem of areas, the area under the moment curve between two points on a beam is proportional to the change in slope between these points, and the moment of this area about any point is proportional to the deflection relative to that point. This method is particularly useful for determining the slope and deflection at specific points along the beam. The application of the moment-area method involves calculating the area under the moment diagram (which represents the bending moment along the length of the beam) and then using this area to determine the change in slope and the deflection at points of interest. The equations governing the moment-area method are given by $\theta = \frac{1}{EI}\int M dx$ for the rotation (slope change) and $\Delta = \frac{1}{EI}\int x M dx$ for the deflection, where E is the modulus of elasticity of the material, I is the moment of inertia of the beam's cross-section, M is the bending moment, x is the distance along the beam, θ is the angle of rotation, and Δ is the deflection.

The conjugate beam method is an ingenious approach that simplifies the calculation of deflections in beams. This method involves transforming the actual beam into a fictitious "conjugate" beam with the same length but subject to modified loading conditions that represent the moment diagram of the original beam. The supports of the conjugate beam are chosen to correspond to the degrees of freedom of the actual beam: a roller support in the conjugate beam represents a free end in the actual beam, a hinge in the conjugate beam corresponds to a roller support in the actual beam, and a free end in the conjugate beam represents a fixed end in the actual beam. The deflection of the actual beam is then directly related to the shear forces in the conjugate beam, and the slope of the actual beam is related to the bending moment in the conjugate beam. The fundamental principle behind the conjugate beam method is that the bending moment at a point in the conjugate beam represents the slope at the corresponding point in the actual beam, and the shear force at a point in the conjugate beam represents the deflection at the corresponding point in the actual beam. The equations used in the conjugate beam method are derived from the basic relationships of shear force, bending moment, and deflection, with the deflection Δ at any point being given by the shear force V in the conjugate beam at that point, and the slope θ at any point being given by the bending moment M in the conjugate beam at that point, with appropriate scaling factors to account for the beam's material and geometric properties.

Each of these methods—double integration, moment-area, and conjugate beam—provides a powerful tool for the analysis of beam and truss deflection, enabling engineers to accurately predict the behavior of structural elements under load. The choice of method depends on the specific requirements of the analysis, including the complexity of the beam or truss configuration, the type and distribution of loads, and the desired information (whether it be the deflection at a particular point, the maximum deflection, or the deflection profile along the entire length of the beam). Mastery of these methods is essential for the structural engineer, providing the analytical foundation necessary to ensure the safety, reliability, and performance of structural systems. Through diligent application of these techniques, engineers can effectively address the challenges posed by diverse loading conditions and complex structural geometries, ensuring that their designs meet the stringent demands of modern engineering practice.

Frame Deflection Analysis

In the realm of structural engineering, understanding the deflection of rigid frames under various loading conditions is paramount. The deflection analysis of rigid frames not only involves the calculation of the magnitude of deflection at specific points but also requires a comprehensive understanding of how different support conditions affect the overall behavior of the structure.

Rigid frames are structures composed of multiple members, where the connection between members is assumed to be moment-resisting. This assumption implies that the members are capable of transferring and resisting bending moments, which significantly influences the deflection characteristics of the frame. The analysis begins with the application of the **principle of superposition**, which allows for the simplification of complex loading conditions by treating each load separately and then combining the results.

The **moment-area method** is a powerful tool in the deflection analysis of rigid frames. This method involves calculating the area under the moment diagram between two points, which directly correlates to the change in slope between those points. The change in slope, integrated over the length of the beam or frame member, gives the deflection. However, the application of this method to rigid frames requires careful consideration of the continuity conditions at the joints and the effect of support settlements.

Support conditions play a critical role in frame deflection. A frame supported on fixed supports will exhibit different deflection characteristics compared to a frame with pinned or roller supports. Fixed supports restrain both translation and rotation, leading to higher moment resistance but potentially larger deflection under certain loading conditions due to the fixed-end moments. In contrast, pinned supports allow rotation, which can lead to reduced moment resistance but also potentially smaller deflections, as the frame can deform more freely to distribute the loads.

The **slope-deflection method** and the **matrix method of analysis** (stiffness method) are also extensively used for analyzing frame deflection. These methods account for the stiffness of each member and the connectivity between members, providing a system of equations that can be solved for joint rotations and displacements. The slope-deflection method explicitly considers the relationship between moments and rotations at the ends of each member, incorporating the effects of fixed and free end conditions. The matrix method, on the other hand, offers a more generalized approach, suitable for computer-aided analysis of complex frames with various support conditions.

The effect of support conditions on frame deflection is not only limited to the immediate response under applied loads but also includes long-term effects such as creep and shrinkage in concrete frames, or thermal expansion and contraction in steel frames. These effects can alter the stress distribution and, consequently, the deflection pattern over time. Therefore, it is crucial to consider both the immediate and long-term deflection behavior when analyzing rigid frames.

The deflection analysis of rigid frames is a multifaceted problem that requires a deep understanding of structural behavior, material properties, and the influence of support conditions. By meticulously applying principles of structural analysis and considering the unique characteristics of each frame, engineers can accurately predict deflection, ensuring the safety, serviceability, and longevity of the structure.

Column Analysis

Buckling Basics: Euler's Formula and Critical Load

Euler's formula for buckling provides a critical tool for understanding the stability of columns under axial compressive loads. The formula is given by

$$P_{cr} = \frac{\pi^2 EI}{(KL)^2}$$

, where P_{cr} is the critical load at which buckling occurs, E is the modulus of elasticity of the material, I is the moment of inertia of the cross-section about the axis of bending, L is the effective length of the column, and K is the column effective length factor, which depends on the conditions of end support. The effective length factor, K, varies with different end conditions, being 1.0 for both ends pinned, 0.5 for both ends fixed, and greater than 1.0 for one end fixed and the other free, indicating how the boundary conditions affect the column's susceptibility to buckling.

The critical load P_{cr} represents the maximum axial load that a column can carry before it becomes unstable due to buckling. It is essential to note that the critical load is directly proportional to the modulus of elasticity E and the moment of inertia I, and inversely proportional to the square of the effective length L. This relationship highlights the importance of the column's geometry and material properties in its stability. A higher moment of inertia, indicating a larger cross-sectional area or a cross-section shape that is more resistant to bending, will increase the critical load, enhancing the column's stability against buckling.

The influence of the column's length on its stability is significant. As the effective length L increases, the critical load P_{cr} decreases, making longer columns more prone to buckling under smaller axial loads. This effect underscores the critical role of selecting appropriate column sizes and shapes in structural design to prevent buckling failures.

Furthermore, the modulus of elasticity E of the material indicates its stiffness. Materials with a higher modulus of elasticity are stiffer and more resistant to deformation, including buckling. Therefore, the choice of material significantly impacts the column's critical load and overall stability.

In practical applications, engineers must consider these factors when designing columns to ensure they can withstand the expected loads without buckling. The selection of material with an appropriate modulus of elasticity, the design of the cross-sectional shape to maximize the moment of inertia, and the careful consideration of the column's length and end conditions are all crucial steps in preventing buckling and ensuring the structural integrity of buildings and other constructions.

The application of Euler's formula extends beyond simple straight columns to include various structural elements where axial compression and stability are concerns. By understanding and applying Euler's formula, engineers can design safer, more efficient structures capable of resisting buckling, thus safeguarding against potential failures that could lead to catastrophic consequences.

Boundary Conditions and Buckling Behavior

Boundary conditions significantly influence the buckling behavior of columns, dictating how these structural elements respond under axial compressive loads. The stability and critical load capacity of a column are directly affected by the constraints imposed at its ends, which can be categorized into three primary types: fixed, pinned, and free-end conditions. Each of these boundary conditions alters the effective length factor (K), which plays a crucial role in determining the buckling load of the column as per Euler's formula $\left(P_{cr} = \dfrac{\pi^2 EI}{(KL)^2} \right)$.

Fixed boundary conditions, where both ends of the column are restrained against rotation, significantly increase the column's resistance to buckling. In this scenario, the effective length factor (K) is reduced to 0.5, effectively halving the effective length (L) of the column and, as a result, quadrupling its critical buckling load compared to a column with both ends pinned. This condition is ideal for maximizing the load-bearing capacity of the column, making it a common choice in the design of rigid frames and other structures where stability under compressive loads is paramount.

Pinned boundary conditions, characterized by ends that are free to rotate but not translate, represent the standard case in Euler's buckling formula, with K set to 1. This condition is commonly found in structural applications where some degree of rotational freedom is necessary or unavoidable. The critical load for a column with pinned ends is lower than that of a fixed column, as the lack of rotational restraint increases the effective length (L) of the column, thereby reducing its capacity to resist buckling.

Free-end conditions, where one end of the column is completely unrestrained, present the most vulnerable scenario for buckling. In this case, the effective length factor (K) increases beyond 1, reflecting the column's heightened susceptibility to buckling due to the lack of any end restraint. This condition is rarely intentional in design due to its instability, but it may occur in scenarios where a column or beam extends beyond its last point of support.

The interplay between the modulus of elasticity (E), moment of inertia (I), and the effective length (L) of the column, modulated by the boundary conditions, underscores the complexity of designing structures that are both efficient and safe under axial loads. Engineers must carefully consider these factors, along with the specific requirements and constraints of each project, to select the most appropriate boundary conditions for columns and other compressive members.

Moreover, the analysis of buckling behavior extends into the selection of materials and cross-sectional shapes that optimize the column's resistance to buckling. Advanced materials with higher moduli of elasticity and cross-sections that maximize the moment of inertia can significantly enhance the stability of columns, even under less favorable boundary conditions. This holistic approach to design, which integrates considerations of material properties, cross-sectional geometry, and boundary conditions, is essential for the development of structures that meet the demands of modern engineering challenges.

In the realm of structural engineering, the meticulous analysis of boundary conditions and their impact on buckling behavior is not merely an academic exercise but a fundamental aspect of ensuring the safety, reliability, and efficiency of built environments. Through the application of principles such as Euler's formula and the careful consideration of boundary conditions, engineers can design columns and other structural elements that stand firm against the forces of nature and time, safeguarding the lives and investments of those who rely on their expertise.

Structural Determinacy and Stability Analysis

Determinate Structures: Criteria and Analysis

Determinacy in structural engineering is a fundamental concept that dictates whether the internal forces and reactions in a structure can be determined solely from static equilibrium equations. For a structure to be considered statically determinate, the number of unknown reactions must exactly equal the number of available equilibrium equations. The criteria for determinacy serve as the foundation for analyzing beams, trusses, and frames, ensuring that each component's behavior under load can be accurately predicted without the need for additional compatibility equations.

Beams are horizontal structural elements designed to carry vertical loads, moments, or both. The determinacy of a beam is assessed by examining the support conditions and the number of reactions they introduce. A simply supported beam, for example, has two reactions at its supports and can be analyzed using the three equilibrium equations available in two dimensions: $\sum F_x = 0$, $\sum F_y = 0$, and $\sum M = 0$. This makes the simply supported beam statically determinate, as the number of unknown reactions does not exceed the number of equilibrium equations.

Trusses are structures composed of members connected at their ends to form a series of triangles. The determinacy of a truss is evaluated using the formula $m + r = 2j$, where m is the number of members, r is the number of reactions, and j is the number of joints. This equation, known as the m-r-j equation, ensures that the truss is statically determinate if it satisfies the condition exactly. If $m + r > 2j$, the truss is statically indeterminate, and additional methods must be employed to analyze it.

Frames are structures that include at least one multi-force member (a member subjected to more than two forces) and can be rigid or pin-connected. The determinacy of a frame can be more complex to ascertain due to the presence of both external and internal reactions. A frame is considered statically determinate if it can be analyzed to find all unknown forces using only the equations of static equilibrium, without requiring additional compatibility conditions. The analysis often involves breaking down the frame into its constituent members and applying the equilibrium equations to each member individually.

Support/Reaction Analysis is crucial in the determinacy assessment and overall structural analysis. The type of support (fixed, pinned, roller, etc.) directly influences the number of reactions and, consequently, the determinacy of the structure. Fixed supports contribute three reactions in two-dimensional structures (a vertical force, a horizontal force, and a moment), while pinned and roller supports contribute fewer reactions (one vertical force for a roller and one vertical and one horizontal force for a pinned support). By accurately identifying and counting the support reactions, engineers can apply the equilibrium equations to solve for unknown forces and moments, ensuring the structure's stability and integrity.

The analysis of determinate structures simplifies the design and evaluation process, allowing engineers to apply fundamental principles of statics to predict structural behavior accurately. Understanding the criteria for determinacy and conducting thorough support/reaction analysis are essential skills for civil engineers, enabling them to design safe and efficient structures that meet the demands of their intended applications.

Stability Analysis

Stability analysis in structural engineering is a critical process that ensures a structure can withstand both the applied loads and the environmental forces it may encounter throughout its lifespan without experiencing failure due to buckling or collapse. This analysis is particularly important for structures that are statically indeterminate, where internal stresses and reactions cannot be determined by static equilibrium equations alone. In these cases, stability checks, the incorporation of redundant supports, and ensuring structural equilibrium are essential strategies to mitigate the risk of instability.

Stability Checks involve the assessment of a structure's ability to remain in equilibrium under expected loads. This includes evaluating the potential for buckling in columns and beams, which is particularly critical in slender structures where the slenderness ratio is high. The Euler's buckling formula,

$$P_{cr} = \frac{\pi^2 EI}{(KL)^2}$$

, provides a basis for calculating the critical load at which buckling occurs, but in real-world applications, factors such as imperfections in the material, geometric nonlinearity, and the effects of inelastic behavior must also be considered. Advanced analysis methods, such as the finite element method (FEM), allow for a more nuanced understanding of how a structure behaves under load, accounting for these complexities and providing a more accurate prediction of stability.

Redundant Supports play a crucial role in enhancing the stability of a structure. Redundancy in structural design refers to the inclusion of additional supports or load paths that are not strictly necessary for maintaining equilibrium under normal conditions but can provide alternative paths for load transfer if a primary load-bearing element fails. This concept is a key aspect of robust design, ensuring that a local failure does not lead to a progressive collapse of the entire structure. For example, in a truss structure, adding redundant members can prevent collapse if one of the members fails, by redistributing the loads among the remaining members. However, the design of redundant supports must be approached with care, as unnecessary redundancy can lead to increased costs and complexity in construction and maintenance.

Structural Equilibrium is the state in which the sum of forces and moments acting on a structure is zero, indicating that the structure is stable and not subject to uncontrolled movement or deformation. Ensuring structural equilibrium involves a detailed analysis of the forces acting on each component of the structure, including both the external loads (such as weights, wind, and seismic forces) and the internal forces generated within the structure in response to these loads. For statically determinate structures, this analysis is straightforward, as the internal forces can be directly calculated from the external loads. However, for statically indeterminate structures, additional methods, such as the method of superposition, compatibility equations, or numerical analysis techniques, must be employed to ensure that all parts of the structure are in equilibrium.

The application of these principles requires a thorough understanding of both the theoretical aspects of structural mechanics and the practical considerations of structural design and construction. Engineers must consider a wide range of factors, including the material properties, the geometry of the structure, the nature of the loads, and the environmental conditions to which the structure will be exposed. By conducting comprehensive stability analyses, incorporating appropriate levels of redundancy, and ensuring that all components of the structure are in equilibrium, engineers can design structures that are both efficient and safe, capable of serving their intended functions while withstanding the forces they will encounter over their lifetimes.

Elementary Statically Indeterminate Structures

Indeterminate Beams and Frames: Superposition & Compatibility

In addressing the challenge of analyzing indeterminate beams and frames, engineers often turn to the principle of superposition and compatibility equations as fundamental tools. These methods allow for the calculation of internal forces and moments in structures that exceed the simple determinacy conditions, where the number of reactions exceeds the number of equilibrium equations available. The complexity of indeterminate structures necessitates a more nuanced approach to ensure accuracy and reliability in structural analysis and design.

The principle of superposition is predicated on the linearity of the equations governing structural behavior. It asserts that the response caused by two or more loads acting on a structure is equivalent to the sum of the responses that would be caused by each load acting individually. This principle is particularly useful in the analysis of statically indeterminate beams and frames, as it allows for the decomposition of complex loading scenarios into simpler, more manageable components. For instance, a beam subjected to a uniform load and a point load can be analyzed by considering the effect of each load separately and then summing the results to obtain the overall response.

Compatibility equations, on the other hand, are rooted in the physical constraints of the structure. They ensure that the deformation of the structure under load is consistent with the continuity and boundary conditions imposed by its geometry and supports. In the context of indeterminate beams and frames, compatibility equations are used to relate the displacements and rotations at various points in the structure to the internal forces and moments. This is essential for determining the additional equations needed to solve for the unknowns in indeterminate systems.

To illustrate, consider an indeterminate beam supported by a fixed support at one end and a roller support at the other, with an intermediate hinge. The analysis would begin by applying the principle of superposition to separate the effects of the external loads. Next, compatibility equations would be formulated to express the continuity of the beam's deflection across the hinge. This might involve setting the deflection at the hinge due to bending moments on either side of it to be equal, thereby ensuring the physical plausibility of the assumed deformation pattern.

The mathematical formulation of these concepts often involves the use of differential equations to describe the beam's deflection curve, $y(x)$, under various loading conditions. The bending moment, $M(x)$, in any segment of the beam can be related to the deflection by the differential equation $EI\dfrac{d^2y}{dx^2} = M(x)$, where E is the modulus of elasticity of the material and I is the moment of inertia of the beam's cross-section about the neutral axis. By integrating this equation, expressions for the deflection and slope at key points can be obtained, which can then be used in the compatibility equations to solve for the unknown reactions and internal forces.

In applying these methods, it is crucial to carefully consider the assumptions underlying linear elastic behavior and the conditions of applicability for the principle of superposition. The analysis must also account for the specific characteristics of the materials and cross-sectional geometry of the beams and frames under consideration. Through meticulous application of superposition and compatibility equations, engineers can achieve a comprehensive understanding of the internal force distributions and deformation patterns in indeterminate structures, thereby ensuring their safety, functionality, and efficiency.

Truss Redundancy Analysis

Truss redundancy is a critical aspect of structural engineering, particularly when analyzing indeterminate trusses. This concept involves the presence of extra members or connections within a truss system that are not necessary for maintaining the structural integrity under nominal load conditions. However, these redundant members play a pivotal role in enhancing the resilience and reliability of the structure, especially in the face of unforeseen loads or failures in other components. The analysis of such trusses requires a detailed understanding of the redistribution of internal forces that occurs as a result of this redundancy.

In indeterminate trusses, the static equilibrium equations alone are insufficient to solve for all the internal forces due to the additional members. Therefore, compatibility conditions must be employed alongside the equilibrium equations to ensure a solvable system. The compatibility conditions are based on the deformation characteristics of the truss, ensuring that the geometry of the structure remains consistent under load. This involves calculating the displacements of joints and comparing the elongation or compression of members that would result from the applied loads, taking into account the material properties and cross-sectional areas.

The method of sections and the method of joints, fundamental to truss analysis, are applied in a modified form to account for the redundancy. For instance, in a redundant truss, certain members may carry no load under a specific loading condition, known as zero-force members. Identifying these members through careful analysis can simplify the problem significantly. However, when dealing with redundancy, it is crucial to consider the load paths and how they are affected by the presence of additional members. The load path describes the direction in which forces are transferred through the structure, and redundancy can alter these paths, leading to a redistribution of forces.

The mathematical representation of this redistribution involves setting up a system of equations that includes both the equilibrium conditions and the compatibility criteria. For example, if a truss has n members and j joints, the general form of the equilibrium equations can be represented as $2j$ equations (considering both horizontal and vertical force balances at each joint). However, in a redundant truss, there will be more unknown member forces than these equations can determine. The compatibility conditions, which relate the deformations of the truss members to the applied loads, provide the additional equations needed. These deformations are often calculated using the formula $\delta = \frac{FL}{AE}$, where δ is the deformation, F is the force in the member, L is the length of the member, A is the cross-sectional area, and E is the modulus of elasticity.

Solving the combined set of equilibrium and compatibility equations typically requires numerical methods, such as matrix algebra or finite element analysis, especially for complex trusses with high degrees of redundancy. The solution will yield the internal forces in all members of the truss, including those that are redundant. It is important to note that the presence of redundancy can lead to indeterminate systems that have multiple valid solutions, each corresponding to different distributions of internal forces but all satisfying the overall equilibrium and compatibility requirements.

In practical applications, the analysis of redundant trusses is essential for ensuring that structures can withstand unexpected loads and failure scenarios without catastrophic collapse. This analysis helps in identifying critical members whose failure could lead to significant redistribution of forces and potentially overstress other components. By understanding the behavior of redundant trusses under various loading conditions, engineers can design safer, more resilient structures that are capable of enduring a wide range of operational and environmental challenges.

Loads, Load Combinations, and Load Paths

Load Types in Structural Systems

Understanding the various types of loads is crucial for the structural analysis and design of buildings and other structures. These loads are categorized based on their origin, duration, and variability. The primary load types include **dead loads**, **live loads**, **lateral loads**, and **moving loads**. Each of these plays a significant role in the overall stability and integrity of a structure.

Dead loads refer to the static forces that are relatively constant over time, including the weight of the structure itself and any permanent fixtures. The calculation of dead loads is straightforward, involving the mass of the materials and the gravitational force. The formula for dead load (D) can be expressed as $D = m \cdot g$, where m is the mass of the structure or component, and g is the acceleration due to gravity. It's essential to accurately estimate dead loads as they provide the baseline forces acting on a structure.

Live loads are variable or transient forces that a structure must support in addition to the dead loads. These can include the weight of people, furniture, vehicles, and snow. Unlike dead loads, live loads can change in magnitude and location, making their prediction more complex. Building codes typically specify minimum live load requirements based on the structure's intended use. Engineers must consider the maximum expected live load during the design process to ensure safety and compliance.

Lateral loads are forces that act horizontally on a structure, such as wind or seismic activity. These loads can induce bending moments, shear forces, and other stressors not accounted for by vertical loads alone. The analysis of lateral loads is critical in areas prone to high winds or earthquakes. For wind loads, the force (F_w) can be estimated using the equation $F_w = 0.5 \cdot \rho \cdot A \cdot V^2$, where ρ is the air density, A is the projected area of the structure, and V is the wind velocity. Seismic loads, on the other hand, depend on the structure's mass, stiffness, and the ground acceleration expected at the site.

Moving loads are those that change position or direction over time within or upon the structure. Examples include vehicles on a bridge, cranes moving along rails, and elevators in a building. The analysis of moving loads is complex, as it must account for the worst-case scenarios of load positioning to ensure the structure can withstand the most critical conditions. Influence lines are instrumental in analyzing the effects of moving loads on structures. An influence line represents the variation in response (such as bending moment or shear force) at a specific point in a structure as a unit load moves across it. The maximum effect of a moving load is determined by positioning the load at points where the influence line reaches its extremities.

Load Combinations and Design Codes

The concept of load combinations is pivotal in structural engineering, ensuring that structures can withstand various types of loads that occur simultaneously or sequentially over their lifespan. The American Society of Civil Engineers (ASCE) provides guidelines for load combinations in ASCE 7, which is widely adopted in design codes such as the International Building Code (IBC). These guidelines stipulate how different loads should be combined to assess the structural integrity under multiple load scenarios, including dead loads (D), live loads (L), snow loads (S), wind loads (W), and earthquake loads (E).

The principle behind load combinations is to account for the fact that not all loads will act at their maximum intensity at the same time. For instance, a heavy snow load is unlikely to coincide with a maximum live load. Therefore, load combinations apply factors to each type of load to simulate realistic scenarios. A common load combination for strength design is $1.2D + 1.6L + 0.5(L_r \, or \, S \, or \, R)$, where L_r is the roof live load, S is the snow load, and R is the rain load. This combination ensures that the structure can support all relevant loads with an appropriate margin of safety.

Load paths play a crucial role in the analysis of load combinations. A load path is the route through which loads are transferred from their point of application to the foundation. Understanding the load path is essential for determining how loads are distributed throughout the structure and for identifying critical points that may be susceptible to failure. The concept of tributary areas is integral to this analysis. The tributary area is the portion of a structure that contributes load to a particular component, such as a beam or column. By calculating the tributary area for each component, engineers can determine the load that each component must support.

The design of a structure must ensure that the load path is continuous and that all loads are effectively transferred to the ground. Discontinuities in the load path can lead to localized failures and compromise the structural integrity. Therefore, during the design process, engineers must meticulously analyze the load path for every load combination to ensure that the structure can safely transfer loads under all possible scenarios.

In combining loads according to design codes, it is also necessary to consider the dynamic effects of certain loads, such as wind and earthquake forces. These dynamic loads introduce lateral forces that can cause swaying and potentially lead to resonance or amplification of the forces. The design must account for these effects by incorporating elements that provide lateral stability, such as shear walls and bracing.

The analysis of load combinations and load paths requires a thorough understanding of structural behavior and the interaction between different types of loads. It involves complex calculations and the use of advanced modeling software to simulate the response of a structure under various load conditions. The goal is to design a structure that is not only efficient and economical but also safe and capable of withstanding the demands placed upon it throughout its service life.

The application of load combinations according to design codes ensures that structures are designed with a comprehensive understanding of all potential loading scenarios. This approach enhances the safety and reliability of structures, protecting both the investment in the built environment and the lives of those who occupy and use these spaces. By meticulously analyzing load paths and tributary areas, engineers can optimize the design to efficiently support and distribute loads, thereby achieving structures that meet the highest standards of safety and performance.

Design of Steel Components

Steel Design Principles

In the realm of structural engineering, particularly when focusing on the design of steel components, it is imperative to adhere to established design philosophies, codes, and incorporate appropriate safety factors. These elements form the backbone of structural integrity, ensuring that steel beams, columns, and tension members can withstand applied loads and stresses over their service life while providing a margin for unforeseen circumstances.

Design philosophies in steel construction are guided by two primary approaches: Allowable Stress Design (ASD) and Load and Resistance Factor Design (LRFD). ASD, a traditional method, operates on the principle that the actual stresses in the structure due to service loads do not exceed the allowable stress, which is a fraction of the yield stress. The formula for ASD can be expressed as $\sigma_{actual} \leq \frac{\sigma_{allowable}}{\Omega}$, where σ_{actual} is the actual stress, $\sigma_{allowable}$ is the allowable stress, and Ω is the safety factor. This approach ensures that under normal conditions, the structure will perform reliably without yielding.

Conversely, LRFD is a more contemporary approach that accounts for both the variability in loads and material strengths by applying different load and resistance factors to each. The fundamental equation for LRFD is $\Phi R_n \geq \gamma_i Q_i$, where R_n is the nominal resistance, Φ is the resistance factor, Q_i represents the effect of various loads, and γ_i are the load factors. LRFD provides a more uniform level of reliability under different loading scenarios by considering the probability of failure in its design criteria.

The adoption of either ASD or LRFD in the design of steel components is dictated by the relevant codes and standards. In the United States, the American Institute of Steel Construction (AISC) publishes the AISC Steel Construction Manual, which provides detailed specifications, codes, and guidelines for both ASD and LRFD methods. Engineers must familiarize themselves with the latest edition of this manual, as it is integral to the design, fabrication, and erection of steel structures.

Safety factors are another critical aspect of steel design, serving as a buffer against uncertainties in the design process, including variations in material properties, inaccuracies in load estimations, and unforeseen environmental impacts. Safety factors are inherently built into the ASD and LRFD methodologies through the allowable stress and resistance factors, respectively. These factors are determined based on statistical analyses of material properties, historical performance data, and considerations of the consequences of failure. For instance, a higher safety factor may be applied to a component whose failure could result in significant economic loss or endanger human life.

In the design of beams, columns, and tension members, engineers must also consider the local and global stability of the structure, which includes buckling and lateral torsional buckling. The slenderness ratio of columns and the unbraced length of beams are critical parameters that influence their design. For tension members, the primary concern is ensuring that the member can carry the applied load without exceeding its yield strength or causing excessive elongation.

The design of steel components is a meticulous process that requires a deep understanding of structural behavior, material science, and the application of design codes and standards. By integrating design philosophies with the appropriate use of safety factors and adherence to codes, engineers can create steel structures that not only meet the required performance criteria but also contribute to the safety, sustainability, and resilience of the built environment.

Steel Connections: Bolted and Welded Joints

In the realm of structural engineering, understanding the intricacies of **steel connections** is paramount for the design and integrity of steel components. Steel connections can be broadly categorized into **bolted** and **welded** connections, each with its unique characteristics, applications, and considerations for load transfer and joint detailing.

Bolted connections are widely used due to their ease of assembly and the ability to disassemble and reassemble if necessary. The design of bolted connections involves considering the bolt strength, the shear and tensile forces acting on the bolt, and the bearing strength of the connected parts. The most common types of bolts used in structural engineering include high-strength bolts, which are typically tension-controlled or slip-critical. The design criteria must ensure that the bolted connection can withstand the forces without failure due to shear, tension, or bearing. The calculation for the design strength of a bolt in shear can be expressed as $F_v = \phi A_b F_u$, where F_v is the shear strength of the bolt, ϕ is the resistance factor, A_b is the nominal unthreaded body area of the bolt, and F_u is the ultimate tensile strength of the bolt material.

Welded connections, on the other hand, provide a continuous connection between steel members, offering superior strength and stiffness compared to bolted connections. The design of welded connections requires a thorough understanding of the welding process, the properties of the weld material, and the effects of welding on the steel members. The strength of a welded connection is determined by the throat thickness of the weld and the strength of the weld material. The effective throat thickness of a fillet weld, which is the most common type of weld used in structural applications, can be calculated as $t = 0.707 \times w$, where w is the leg size of the fillet weld. The design strength of the weld is then determined by multiplying the effective throat thickness by the length of the weld and the weld material strength.

Joint detailing is crucial for both bolted and welded connections to ensure proper force transfer and to avoid stress concentrations that could lead to failure. For bolted connections, detailing involves specifying the bolt size, type, spacing, edge distances, and arrangement to ensure even distribution of forces and to prevent bolt failure. For welded connections, detailing includes specifying the type of weld, size, length, and location to ensure that the weld is capable of transferring the forces without failure.

In both bolted and welded connections, **load transfer** mechanisms are essential for the stability and integrity of the structure. Bolted connections transfer load through shear in the bolts and bearing on the bolt holes, while welded connections transfer load through the weld material itself. The design and detailing of steel connections must ensure that the load path is clear and direct, and that the connections have sufficient strength and stiffness to transfer the loads safely.

Design of Reinforced Concrete Components

Concrete Design Basics

Moving forward from the foundational understanding of steel connections, the design of reinforced concrete components, specifically beams and slabs, requires a deep dive into the principles that govern their structural integrity and functionality. The design of these components is governed by various codes and standards, which ensure safety, durability, and efficiency in construction practices. The American Concrete Institute (ACI) provides comprehensive guidelines through ACI 318, "Building Code Requirements for Structural Concrete," which is a critical resource for engineers. This code outlines the minimum requirements for the design and construction of structural concrete elements, incorporating the latest research and technological advancements in the field.

The design of reinforced concrete beams and slabs begins with an understanding of the material properties of concrete and steel, including their stress-strain behavior under applied loads. Concrete, being strong in compression but weak in tension, is reinforced with steel bars (rebar) to enhance its tensile strength. The compatibility of deformation between concrete and steel under load is a fundamental concept, ensuring that both materials act together as a single unit. The design process involves determining the required amount of reinforcement to resist bending moments, shear forces, and torsional moments, ensuring that the structure can support the imposed loads without failure.

Safety factors, or factors of safety (FOS), play a crucial role in the design of reinforced concrete structures. These are incorporated to account for uncertainties in material properties, load assumptions, and variations in construction practices. The ACI code specifies minimum safety factors for different failure modes, including flexural failure, shear failure, and bond failure, among others. For instance, the design of a reinforced concrete beam for flexural strength involves calculating the nominal moment capacity (M_n) and comparing it with the factored moment (M_u), which includes a safety margin above the expected maximum moment due to loads on the beam. The equation $\phi M_n \geq M_u$ must be satisfied, where ϕ is the strength reduction factor provided by ACI 318, which varies depending on the type of stress and the failure mode considered.

In addition to the strength requirements, serviceability criteria such as deflection control and crack width limitation are essential in the design of beams and slabs. These criteria ensure that the structure not only remains safe under load but also performs adequately for its intended use without causing discomfort or damage to finishings and non-structural elements. The ACI code provides limits on the maximum allowable deflection and methods to calculate deflection, taking into account the effects of creep, shrinkage, and temperature changes on long-term deflection.

The design of slabs, whether one-way or two-way, follows similar principles but with additional considerations for the distribution of loads and the support conditions. One-way slabs are supported on two opposite sides and carry loads primarily in one direction, while two-way slabs are supported on all sides and distribute loads in two directions. The design of two-way slabs often involves more complex analysis methods, such as the direct design method or the equivalent frame method, to accurately model the load distribution and determine the required reinforcement.

In summary, the design of reinforced concrete beams and slabs is a multifaceted process that integrates material properties, structural analysis, and code requirements to achieve safe, functional, and efficient structural components. The adherence to established codes and the application of safety factors ensure that these components can withstand the imposed loads and environmental conditions they will face during their service life.

Column and Member Design

Transitioning from the foundational principles of beam and slab design, the focus shifts to the critical aspects of column and member design in reinforced concrete structures. The design of reinforced concrete columns is governed by the need to resist axial loads, bending moments, and shear forces, which are prevalent in most structural frameworks. The American Concrete Institute's ACI 318 provides the necessary guidelines and formulas for the design and detailing of these elements, ensuring their structural integrity and safety under applied loads.

The design process for reinforced concrete columns begins with the determination of interaction diagrams, which represent the combination of axial load and bending moment that the column can resist. These diagrams are essential for understanding the column's capacity and are derived from the material properties of concrete and steel, as well as the geometric characteristics of the column cross-section. The strength of a column is significantly influenced by its slenderness ratio, which is the ratio of its effective length to its least radius of gyration. Slender columns are more susceptible to buckling under axial loads, necessitating additional design considerations to mitigate the risk of instability.

ACI 318 specifies the use of the $P - \Delta$ effect for slender columns, which accounts for the additional moment induced by axial load acting on an out-of-plumb column. The design must ensure that the amplified moment, $M_2 = M_1 + \Delta M$, does not exceed the column's moment capacity, where M_1 is the moment due to applied loads and ΔM is the moment induced by the column's displacement. The calculation of ΔM involves the column's stiffness, the applied axial load, and the initial displacement or out-of-plumbness.

Detailing of reinforcement in columns is another critical aspect that directly impacts the structural performance and ductility of the column. The ACI code mandates the provision of longitudinal reinforcement to resist tensile stresses and lateral ties or spirals to confine the core of the column, enhancing its ductility and shear resistance. The spacing, diameter, and configuration of these ties are regulated to ensure adequate confinement and prevent buckling of the longitudinal bars under seismic loads or high axial stresses.

The transition region between beams and columns, commonly referred to as the joint region, requires special attention in detailing to ensure the transfer of forces without significant stress concentrations or failure. The integrity of this region is paramount in seismic design, where the joint must accommodate the inelastic deformations of the beam and column ends without losing its load-carrying capacity. ACI 318 provides specific provisions for the reinforcement detailing in these regions, including criteria for the anchorage of reinforcement and the use of stirrups or ties to resist shear forces.

In addition to the structural considerations, the aesthetic and functional requirements of the column design must also be addressed. The size and shape of the column, along with the surface finish, play a significant role in the architectural expression of a building. Furthermore, the placement of columns should consider the building's use and occupancy, ensuring that the columns do not obstruct the space or impede the functionality of the structure.

The design and detailing of reinforced concrete columns are complex processes that integrate structural engineering principles with practical considerations of construction and use. Adherence to the guidelines provided by ACI 318 and a thorough understanding of the interaction between different forces and the behavior of materials under load are essential for the successful design of safe, efficient, and durable reinforced concrete columns.

Chapter 12: Geotechnical Engineering

Index Properties and Soil Classifications

Soil Properties: Indicators of Soil Behavior

Grain size is a fundamental property of soil that significantly influences its behavior and classification. The distribution of grain sizes within a soil sample, often represented by a grain size distribution curve, provides insight into the soil's permeability, compaction characteristics, and its classification as coarse-grained (sand and gravel) or fine-grained (silt and clay). The Unified Soil Classification System (USCS) and the American Association of State Highway and Transportation Officials (AASHTO) system categorize soils based on their grain size distribution. Coarse-grained soils, with larger particle sizes, generally exhibit higher permeability, allowing for quicker drainage and less susceptibility to frost action compared to fine-grained soils, which are characterized by smaller particle sizes that contribute to higher capillarity and plasticity.

The Atterberg limits, including the liquid limit, plastic limit, and shrinkage limit, are critical in defining the fine-grained soils' behavior under varying moisture conditions. These limits are key to understanding the soil's transition between different states: from solid to plastic and from plastic to liquid. The liquid limit (LL) is the moisture content at which soil transitions from a plastic to a liquid state, indicating its ability to flow. The plastic limit (PL) is the moisture content at which soil transitions from a semi-solid to a plastic state, signifying its moldability. The difference between these two limits, known as the plasticity index ($PI = LL - PL$), provides a measure of the soil's plasticity, which is directly related to its clay content and type. Soils with high plasticity indexes are typically more susceptible to volume changes with moisture content variations, impacting their stability and suitability for construction projects. The shrinkage limit (SL) is the moisture content at which further loss of moisture does not result in a decrease in volume, marking the transition from plastic to solid state. These limits are determined through standardized laboratory tests and are essential for classifying fine-grained soils and predicting their behavior under load and moisture changes.

Specific gravity of soil solids (G_s), a dimensionless quantity, is another vital property that influences soil behavior. It represents the ratio of the mass of a given volume of soil solids to the mass of an equal volume of water at a specified temperature, typically 4°C, where water has its maximum density. The specific gravity values of most mineral soils range between 2.60 and 2.80. This property is crucial for calculating phase relationships, including porosity, void ratio, and degree of saturation, which are fundamental for understanding soil compaction, consolidation, and shear strength. The specific gravity is determined using either a pycnometer, a density bottle, or a gas displacement method, and it plays a significant role in the analysis and design of geotechnical engineering projects by influencing the calculation of stresses within the soil mass.

Understanding the interplay between grain size, Atterberg limits, and specific gravity is essential for geotechnical engineers to predict soil behavior under various conditions, design foundations, and mitigate risks associated with soil-structure interaction. These soil properties, indicative of the soil's physical and mechanical characteristics, form the basis for soil classification systems that guide engineers in selecting appropriate construction techniques and materials management strategies for infrastructure projects. The behavior of soil under external loads, its capacity to support structures, and its response to environmental changes are all influenced by these fundamental properties, underscoring the importance of thorough soil analysis in the planning and execution of engineering projects.

Classification Systems: USCS and AASHTO Soils

The Unified Soil Classification System (USCS) and the American Association of State Highway and Transportation Officials (AASHTO) system are pivotal in categorizing soils, each with a unique approach based on particle size distribution and plasticity characteristics. The USCS, widely utilized in geotechnical engineering, classifies soils into major groups: coarse-grained soils, fine-grained soils, and organic soils, further subdivided based on specific criteria such as grain size, plasticity index, and liquid limit. Coarse-grained soils are identified as having more than 50% of their material by weight larger than a No. 200 sieve, and are classified into gravels (G) and sands (S), with further subdivisions based on their gradation and fines content. Fine-grained soils, with more than 50% passing a No. 200 sieve, are classified into clays (C) and silts

(M), with distinctions made based on the plasticity index and liquid limit. The plasticity chart, a graphical representation plotting the liquid limit and plasticity index, serves as a crucial tool in this classification, enabling the differentiation between low-plasticity silts and clays from high-plasticity ones.

The AASHTO system, primarily used in highway and transportation projects, categorizes soils into eight groups, A-1 through A-8, based on their suitability for road construction, which is determined by particle size distribution, liquid limit, and plasticity index. The A-1 group represents the most desirable materials, such as well-graded sands and gravels with low plasticity, while the A-7 group includes soils with high plasticity, which are less suitable due to their significant volume change with moisture variations. The AASHTO classification includes a group index, a numerical value that quantifies the soil's expected performance as a subgrade material, with higher values indicating poorer performance. The calculation of the group index is based on an empirical formula that incorporates the percentage passing a No. 200 sieve, the liquid limit, and the plasticity index, providing a quantitative measure to compare the quality of soil materials.

Both the USCS and AASHTO systems offer a structured approach to soil classification, yet their application and emphasis differ. The USCS focuses on the engineering properties of soils relevant to foundation design, embankment construction, and other geotechnical engineering projects, emphasizing the soil's behavior under load and its permeability. In contrast, the AASHTO system is tailored towards evaluating soils for their suitability in road construction, concentrating on the stability and drainage characteristics essential for pavement layers. Understanding the distinctions between these classification systems is crucial for engineers to select the appropriate system based on the project requirements, ensuring that the soil's characteristics are accurately represented and effectively utilized in design and construction processes.

The integration of soil classification systems into geotechnical analysis facilitates the prediction of soil behavior under various conditions, aiding in the design of foundations, the assessment of slope stability, and the determination of the need for soil improvement techniques. By categorizing soils based on standardized criteria, engineers can communicate effectively about the material properties, anticipate potential challenges in construction, and devise strategies to mitigate risks associated with soil variability. The choice between the USCS and AASHTO systems, or the combination of both, depends on the specific engineering objectives, the nature of the project, and the regulatory requirements governing the construction activities. Through the application of these classification systems, engineers are equipped with a methodical approach to evaluate soil suitability, optimize material selection, and ensure the structural integrity and longevity of civil engineering projects.

Phase Relations

Basic Soil Relationships

Understanding the basic relationships between void ratio, porosity, moisture content, and degree of saturation is crucial for geotechnical engineering applications. These parameters are fundamental in assessing the physical characteristics of soil, which in turn influence its behavior under various conditions.

Void ratio (e) is a dimensionless quantity that represents the ratio of the volume of voids (V_v) to the volume of solids (V_s) in a soil mass. It is expressed as $e = \dfrac{V_v}{V_s}$. This parameter is essential for understanding the soil's capacity to retain water and its potential for compression and consolidation.

Porosity (n) is another dimensionless quantity, closely related to void ratio, defined as the ratio of the volume of voids to the total volume of the soil (V_t), given by $n = \dfrac{V_v}{V_t} = \dfrac{V_v}{V_v + V_s}$. Porosity is a critical factor in determining the permeability of the soil, affecting how fluids flow through the soil mass.

The relationship between void ratio and porosity can be described by the equation $n = \dfrac{e}{1+e}$, illustrating how these two parameters are interdependent yet distinct measures of the soil's void space.

Moisture content (w) is a measure of the amount of water contained in the soil, expressed as a percentage of the weight of the dry soil (W_d). It is calculated using the formula $w = \dfrac{W_w}{W_d} \times 100\%$, where W_w is the weight of the water in the soil. Moisture content is a vital parameter for understanding the soil's condition and its effect on soil strength, compressibility, and other engineering properties.

Degree of saturation (S_r) is the ratio of the volume of water (V_w) to the volume of voids, expressed as a percentage, $S_r = \dfrac{V_w}{V_v} \times 100\%$. This parameter indicates how much of the soil's pore space is filled with water, providing insights into the soil's hydraulic characteristics and its behavior under loading conditions.

The interrelationships among these parameters are key to analyzing and solving various geotechnical engineering problems, such as calculating settlement, evaluating slope stability, and designing foundations. Understanding these relationships allows engineers to predict how soils will respond to environmental changes, loading, and other external factors, ensuring the safety and stability of civil engineering structures.

Weight-Volume Analysis in Engineering

Weight-volume analysis is a fundamental aspect of geotechnical engineering, providing essential insights into the physical properties of soil and their implications for various engineering applications. This analysis involves the determination of **dry unit weight** (γ_d), **saturated unit weight** (γ_{sat}), and **submerged unit weight** (γ') of soils, each of which plays a critical role in the design and analysis of foundations, earthworks, and other soil-structure interaction problems.

Dry unit weight is defined as the total weight of soil particles per unit volume of soil, excluding the pore water. It is calculated using the formula:

$$\gamma_d = \dfrac{W_d}{V_t}$$

where W_d is the weight of the dry soil, and V_t is the total volume of the soil. Dry unit weight is a critical parameter in evaluating the compactness and shear strength of soils, influencing the stability of slopes and the bearing capacity of foundations.

Saturated unit weight refers to the weight of soil per unit volume when the voids within the soil mass are completely filled with water. It is given by:

$$\gamma_{sat} = \frac{W_d + W_w}{V_t}$$

where W_w is the weight of water filling the soil voids. This parameter is essential for understanding the behavior of soil under conditions of full saturation, such as during the design of retaining structures and the analysis of seepage through earth dams.

Submerged unit weight, also known as the buoyant unit weight, is the effective weight of soil when submerged in water. It accounts for the buoyant effect of water and is calculated as:

$$\gamma' = \gamma_{sat} - \gamma_w$$

where γ_w is the unit weight of water. Submerged unit weight is particularly important in the analysis of soil stability under water, including the design of underwater foundations and the assessment of uplift pressure on submerged structures.

The engineering implications of these weight-volume parameters are vast. For instance, the dry unit weight of soil is used to assess the compaction level of earth fills and embankments, ensuring their stability and durability. The saturated unit weight is crucial for evaluating the potential for liquefaction under seismic loading, a critical consideration in earthquake-prone areas. Meanwhile, the submerged unit weight is instrumental in calculating the effective stress in submerged soils, which influences the stability of slopes and excavations near water bodies.

Understanding the relationships between these weight-volume parameters and their impact on soil behavior under various conditions is essential for geotechnical engineers. It enables the accurate prediction of soil-structure interaction, ensuring the safety, stability, and longevity of civil engineering projects. Through meticulous weight-volume analysis, engineers can make informed decisions regarding soil treatment, foundation design, and the mitigation of geotechnical risks, ultimately contributing to the successful completion of engineering projects that stand the test of time and nature.

Laboratory and Field Tests

Laboratory Tests: Explain compaction, permeability, and shear strength tests performed in controlled settings.

Laboratory tests are essential for evaluating the properties of soil, particularly in terms of compaction, permeability, and shear strength.

The **Proctor Compaction Test** is a widely used method to determine the optimal moisture content and maximum dry density of soil. This test involves compacting soil at various moisture levels to identify the point at which the soil achieves its highest density. The results are crucial for ensuring that soil is adequately compacted in construction projects, which directly affects the stability and load-bearing capacity of structures.

The **Permeability Test** assesses the rate at which water flows through soil, providing insights into its drainage characteristics. This test is vital for understanding how quickly water can move through different soil types, which influences decisions related to site drainage, foundation design, and environmental impact assessments.

The **Direct Shear Test** evaluates the shear strength of soil by applying a parallel force to a soil sample until failure occurs. This test measures the soil's resistance to sliding along a failure plane, which is critical for analyzing slope stability, retaining wall design, and other geotechnical applications.

In contrast, the **Atterberg Limits Test** is used primarily for fine-grained soils to determine their plasticity characteristics, while the **Consolidation Test** measures soil compressibility rather than compaction. These tests, although important, do not directly assess the compaction or shear strength of soil.

Understanding these laboratory tests and their implications is fundamental for civil engineers and geotechnical professionals, as they inform the design and construction processes, ensuring safety and performance in engineering projects.

Field Tests: Discuss SPT, CPT, and vane shear tests for determining in-situ soil properties.

The vane shear test is a crucial method for determining the in-situ shear strength of cohesive soils while minimizing disturbance to the soil structure. This test employs a four-bladed vane that is inserted into the soil at the desired depth. The vane is then rotated, and the torque required to cause failure is measured. This torque is directly related to the shear strength of the soil, allowing for an accurate assessment of its properties.

In contrast, the Standard Penetration Test (SPT) is primarily focused on measuring the resistance of soil to penetration. While it provides valuable information regarding soil density and strength, it does not directly measure shear strength. The SPT involves driving a split-barrel sampler into the ground and counting the number of blows required to penetrate a specific depth, which can be correlated to soil properties but lacks the precision of shear strength measurement.

Similarly, the Cone Penetration Test (CPT) assesses soil stratigraphy and properties by measuring the resistance encountered by a cone as it is pushed into the ground. This test yields valuable data on soil behavior, including friction and pore pressure, but it does not specifically target shear strength as effectively as the vane shear test. Therefore, for applications requiring precise determination of shear strength in cohesive soils, the vane shear test is the preferred method due to its minimal disturbance and direct measurement capabilities.

Effective Stress

Stress Basics in Soil Systems

Total stress, denoted as σ, in soil mechanics, refers to the total force per unit area exerted by a soil mass. This includes both the solid particles and the pore water within the soil. It is calculated as the weight of the soil column above a given point and can be expressed as $\sigma = \gamma \cdot h$, where γ is the unit weight of the soil and h is the height of the soil column above the point of interest. Total stress is a critical parameter in geotechnical engineering, as it influences the design and analysis of foundations, retaining walls, and other soil-structure interaction systems.

Effective stress, on the other hand, is the concept introduced by Karl Terzaghi, often considered the father of soil mechanics, to describe the stress that is effectively carried by the soil skeleton. It is the difference between the total stress and the pore water pressure (u) within the soil. The effective stress, denoted as σ', is given by the equation $\sigma' = \sigma - u$. This concept is fundamental in understanding the behavior of soils under load, as it directly influences the soil's strength, compressibility, and permeability. Effective stress is a key factor in the analysis and design of geotechnical engineering projects, as it determines the soil's capacity to support structures.

Pore water pressure (u) is the pressure exerted by water within the pores of a soil mass. It plays a significant role in the effective stress principle, as it counteracts the stress transmitted through the soil particles. The presence of water in the soil pores can significantly alter the soil's behavior, reducing its effective stress and, consequently, its shear strength. This phenomenon is particularly evident in cases of soil saturation, leading to conditions such as liquefaction in seismic events. Pore water pressure is measured in terms of force per unit area and can vary with changes in external loads, groundwater levels, and soil permeability.

The relationship between total stress, effective stress, and pore water pressure is crucial for predicting soil behavior under various loading conditions. For instance, during the construction of a structure, the increase in total stress due to the added load can lead to an increase in pore water pressure if the drainage is insufficient. This, in turn, reduces the effective stress, potentially compromising the soil's ability to support the structure. Understanding this relationship allows engineers to design appropriate drainage and reinforcement solutions to mitigate risks associated with changes in soil stress states.

The effective stress principle also underpins the analysis of consolidation behavior in soils. When a load is applied to a saturated soil, the immediate response is an increase in pore water pressure, as the water in the soil pores bears the additional load. Over time, as the water gradually dissipates from the soil, the effective stress increases, leading to soil consolidation and settlement. This time-dependent behavior is a critical consideration in the design of foundations and earth structures, requiring careful analysis to ensure long-term stability and performance.

The concepts of total stress, effective stress, and pore water pressure are foundational to geotechnical engineering, providing a framework for understanding and predicting soil behavior under load. These principles guide the design and analysis of geotechnical solutions, ensuring the safety and reliability of civil engineering projects. Through the application of these concepts, engineers can address the challenges posed by soil-structure interactions, optimizing designs to achieve desired performance and stability.

Applications of Effective Stress

The impact of **effective stress** on **settlement**, **shear strength**, and **slope stability** is a cornerstone in geotechnical engineering, directly influencing the design and safety of civil engineering structures. Understanding these applications is crucial for engineers to predict and mitigate potential issues in soil-structure interaction.

Settlement occurs when the soil beneath a structure compresses due to the weight of the structure. The effective stress principle is vital in calculating settlement, as it determines the load that the soil skeleton can bear. An increase in effective stress, typically due to an added structural load, leads to soil particles rearranging into a denser configuration, resulting in settlement. The magnitude of settlement can be estimated using the consolidation theory, which is based on the effective stress concept. The primary consolidation settlement is given by the equation:

$$\Delta H = \frac{C_c \cdot H_0 \cdot \log\left(\frac{\sigma'_f}{\sigma'_0}\right)}{1 + e_0}$$

where ΔH is the change in thickness of the soil layer, C_c is the compression index, H_0 is the original thickness of the layer, σ'_0 and σ'_f are the initial and final effective stresses, and e_0 is the initial void ratio. This equation underscores the direct relationship between effective stress and settlement, highlighting the importance of accurately assessing changes in effective stress to predict settlement behavior.

Shear strength of soil is its capacity to resist shearing stresses and is a critical factor in ensuring the stability of slopes and foundations. The shear strength of soil is primarily governed by its effective stress, as articulated by the Mohr-Coulomb failure criterion:

$$\tau = c' + \sigma' \tan(\varphi')$$

where τ is the shear strength, c' is the cohesion intercept, σ' is the effective normal stress, and φ' is the angle of internal friction. This relationship illustrates that as effective stress increases, the shear strength of the soil also increases, assuming constant soil properties. Therefore, managing water levels and ensuring proper drainage around foundations and slopes are essential to maintain or enhance the effective stress and, consequently, the shear strength of the soil.

Slope stability is significantly influenced by the effective stress within the soil mass forming the slope. The stability of a slope is assessed by evaluating the factor of safety against sliding along potential failure surfaces, which is a function of the shear strength of the soil. The factor of safety (FS) is given by:

$$FS = \frac{\sum \tau}{\sum \tau_f}$$

where $\sum \tau$ is the sum of resisting forces along the failure surface, and $\sum \tau_f$ is the sum of driving forces. Since the shear strength (τ) is dependent on the effective stress, any condition that reduces the effective stress, such as an increase in pore water pressure, can decrease the slope's factor of safety, leading to instability. Thus, controlling groundwater flow and surface water infiltration into slopes is crucial for maintaining stability.

The applications of effective stress in geotechnical engineering are vast and critical for the design and analysis of civil engineering projects. By understanding and applying the principles of effective stress, engineers can ensure the safety, stability, and longevity of structures and earthworks, addressing the challenges posed by soil behavior under various loading conditions.

Stability of Retaining Structures

Earth Pressure Theories: Rankine and Coulomb

The **Rankine theory** of earth pressure is a classical approach that simplifies the analysis by assuming that the soil mass is homogeneous, isotropic, and the failure surface within the soil is planar. This theory distinguishes between active and passive earth pressures. **Active earth pressure** (P_a) occurs when a retaining wall moves away from the soil, allowing the soil mass to expand and mobilize its shear strength along the failure plane. The magnitude of active earth pressure is given by the equation:

$$P_a = \frac{1}{2}\gamma H^2 \left(\cos^2(\phi) - \sqrt{\cos^2(\phi) - \cos^2(\theta)} \right)$$

where γ is the unit weight of the soil, H is the height of the wall, ϕ is the angle of internal friction of the soil, and θ is the angle of the back of the wall with the horizontal. This formula simplifies to a more commonly used form when $\theta = 0$, indicating a vertical wall.

Passive earth pressure (P_P), on the other hand, develops when the wall moves towards the soil, compressing the soil mass and mobilizing resistance against the movement. The passive earth pressure can be calculated as:

$$P_P = \frac{1}{2}\gamma H^2 \left(\cos^2(\phi) + \sqrt{\cos^2(\phi) - \cos^2(\theta)} \right)$$

The **Coulomb theory** provides a more general solution to the problem of earth pressure by considering the effects of wall friction, wall inclination, soil-wall adhesion, and backfill slope. Coulomb's earth pressure theory is expressed through the equation:

$$P = \frac{1}{2}\gamma H^2 K$$

where K is the earth pressure coefficient, which varies depending on whether the condition is active or passive. The coefficient K for active and passive cases can be determined using complex trigonometric relationships that account for the aforementioned factors, making Coulomb's method more versatile for different wall and soil conditions.

Both Rankine and Coulomb theories are essential for the design and analysis of retaining structures, providing engineers with the tools to estimate the lateral earth pressures exerted on retaining walls, sheet piles, and braced cuts. Understanding the distinctions and applications of these theories allows for the optimization of retaining structure designs, ensuring stability while minimizing construction costs.

The selection between Rankine and Coulomb theories in practice depends on the specific conditions of the project, including soil characteristics, wall geometry, and loading conditions. For instance, Rankine's theory is often preferred for simpler cases with vertical walls and horizontal backfill surfaces, due to its straightforward application. Coulomb's theory, with its ability to incorporate a wider range of factors, is suited for more complex scenarios where wall and soil interactions are more nuanced.

In applying these theories, it is crucial to conduct thorough soil investigations to accurately determine the soil parameters (γ, ϕ, and others) that significantly influence the calculated earth pressures. Additionally, the assumptions inherent in each theory must be critically evaluated in the context of each project to ensure the reliability of the analysis.

By integrating these theoretical frameworks with empirical data and modern analytical techniques, geotechnical engineers can effectively design and evaluate the stability of retaining structures, addressing the challenges posed by diverse soil conditions and loading scenarios.

Wall Stability Analysis

In assessing the stability of retaining walls, engineers must meticulously evaluate the potential for **overturning, sliding**, and issues related to **bearing capacity**. These factors are critical in ensuring the safety and functionality of retaining structures under various loading conditions.

Overturning occurs when the moment caused by lateral earth pressure exceeds the stabilizing moments from the wall's weight and any additional resisting forces. The calculation of the overturning moment (M_o) involves integrating the lateral earth pressure distribution along the height of the wall. This is typically represented by the formula $$M_o = \int_0^H p(x) x \, dx$$, where

$p(x)$ is the pressure at depth x, and H is the total height of the wall. To counteract this moment, the stabilizing moment (M_s) must be greater, incorporating the wall's self-weight and any surcharge loads. The safety factor against overturning is given by $FS_o = \frac{M_s}{M_o}$, where values typically greater than 1.5 are considered safe.

Sliding refers to the horizontal displacement of a wall due to insufficient friction or shear resistance at the base. The sliding force (F_s) is primarily the result of the horizontal component of the earth pressure. The resisting force (F_r) is the product of the normal force (N) and the friction coefficient (μ) between the wall base and the foundation soil, expressed as $F_r = \mu N$. The safety factor against sliding is calculated as $FS_s = \frac{F_r}{F_s}$, with acceptable values usually exceeding 1.5 to ensure adequate resistance.

Bearing capacity failure occurs when the soil beneath the wall fails to support the combined load of the wall and the soil behind it, leading to settlement or rotation. The ultimate bearing capacity (q_u) of the soil is a function of its cohesion (c), internal friction angle (ϕ), and the density (γ). The bearing pressure (q) under the wall should not exceed the allowable bearing capacity ($q_a = \frac{q_u}{FS_b}$), where FS_b is the safety factor for bearing capacity, commonly taken as greater than 2.0 for conservative design.

In practical engineering applications, detailed geotechnical investigations provide the necessary parameters to evaluate these stability concerns accurately. Additionally, modern design often incorporates reinforcement, such as geogrids or tiebacks, to enhance the stability of retaining walls against these failure modes. The integration of these reinforcements needs to be carefully designed based on the specific site conditions, wall type, and expected loads to ensure the long-term performance of the retaining structures.

Shear Strength

Shear Strength Fundamentals

The **Mohr-Coulomb failure criterion** is a fundamental model used to describe the shear strength of soils and rocks. This criterion is pivotal in geotechnical engineering for analyzing and designing against failure in soil masses, slopes, and retaining structures. It is expressed mathematically as $\tau = c + \sigma \tan(\phi)$, where τ represents the shear strength of the material, c is the cohesion, σ is the normal stress, and ϕ is the angle of internal friction. The cohesion c is a measure of the soil's ability to withstand shear stress without the presence of any normal stress and is a direct indicator of the soil's inherent stickiness or its ability to adhere to itself. The angle of internal friction ϕ, on the other hand, provides insight into the material's resistance to sliding over itself when under stress.

The **Mohr-Coulomb theory** assumes that failure occurs along a plane where the shear stress τ exceeds the material's shear strength. The theory is graphically represented by the Mohr-Coulomb failure envelope in a plot of shear stress versus normal stress. The slope of this line, given by $\tan(\phi)$, and its intercept with the shear stress axis, c, define the material's shear strength parameters. This graphical representation is crucial for understanding the conditions under which a soil or rock mass may fail and is extensively used in the design and analysis of geotechnical structures.

In practical applications, determining the values of c and ϕ is achieved through laboratory tests such as the triaxial compression test, direct shear test, and unconfined compression test. These tests provide engineers with critical data to model the behavior of soil and rock masses under various loading conditions. For instance, in slope stability analysis, the Mohr-Coulomb failure criterion is used to evaluate the potential for landslide occurrences by assessing the balance between driving forces, which promote sliding, and resisting forces, which oppose it. Similarly, in the design of foundations, the criterion helps in determining the bearing capacity of soils, ensuring that structures are supported adequately to prevent excessive settlement or failure.

The Mohr-Coulomb failure criterion is crucial for geotechnical engineers in predicting and addressing potential failures in soil and rock masses. This criterion underpins the design of safe and effective civil engineering structures, ranging from basic retaining walls to intricate earth dams and slopes. Its straightforward nature and versatility establish it as a fundamental component in geotechnical engineering analyses, offering a dependable framework for evaluating shear strength and maintaining the stability of geotechnical systems.

Tests and Applications: Discuss direct shear, triaxial, and unconfined compression tests for shear strength evaluation.

The unconfined compression test is a crucial method for evaluating the shear strength of cohesive soils. In this test, a cylindrical soil sample is subjected to a vertical axial load without any lateral confinement, allowing the sample to deform freely. This process enables the determination of the unconfined compressive strength, which serves as a direct measure of the soil's shear strength.

In contrast, the direct shear test involves applying a horizontal force to a soil sample that has been split along a predetermined plane. This setup allows for the straightforward measurement of shear strength parameters under a defined normal stress. The test is particularly useful for understanding the shear behavior of soils under various loading conditions.

The triaxial test provides a more comprehensive assessment of soil shear strength by applying both axial and confining pressures to a cylindrical sample. This method allows for the simulation of in-situ conditions, enabling engineers to analyze how soils will behave under different stress states. The triaxial test can be conducted in various configurations, including unconsolidated undrained (UU), consolidated undrained (CU), and consolidated drained (CD) tests, each providing valuable insights into the soil's mechanical properties and stability under load.

Bearing Capacity

Bearing Capacity of Shallow Foundations

The ultimate bearing capacity (q_u) is a critical parameter in the design of shallow foundations, representing the maximum pressure a soil can withstand before failure occurs. This failure is typically characterized by shear failure in the soil beneath the foundation and can be determined through various methods, including empirical formulas, field tests, and numerical analysis. The

Terzaghi bearing capacity equation is one of the most widely used formulas for calculating q_u for strip footings on cohesive soils, given by:

$$q_u = cN_c + \gamma D_f N_q + 0.5\gamma B N_\gamma$$

where c is the cohesion of the soil, γ is the unit weight of the soil, D_f is the depth of the foundation, B is the width of the foundation, and N_c, N_q, and N_γ are bearing capacity factors related to the angle of internal friction (φ) of the soil.

The net bearing capacity (q_n) takes into account the weight of the soil above the foundation level that is displaced by the foundation. It is calculated by subtracting the overburden pressure (γD_f) from the ultimate bearing capacity:

$$q_n = q_u - \gamma D_f$$

This value is essential for ensuring that the foundation does not impose excessive stress on the soil, leading to settlement or failure.

The allowable bearing capacity (q_a) is the maximum pressure that can be safely applied to the soil, taking into account a factor of safety (FS). It is determined by dividing the net bearing capacity by the factor of safety:

$$q_a = \frac{q_n}{FS}$$

The factor of safety is chosen based on the level of risk acceptable for the project and the variability in the soil properties, typically ranging from 2 to 3 for most projects. This conservative approach ensures that the foundation will not fail under the most adverse conditions anticipated.

The determination of these bearing capacity values is fundamental in the design process of shallow foundations, ensuring that the structures supported by these foundations are safe and stable. Accurate assessment of the soil's bearing capacity helps in selecting the appropriate type and dimensions of the foundation, thus preventing structural damage that could result from foundation failure. It is imperative for engineers to consider the variability in soil properties across a site, as well as the impact of environmental factors such as moisture content and frost action, which can significantly affect the soil's bearing capacity. Therefore, geotechnical site investigations, including soil sampling and testing, are crucial steps in obtaining the necessary data for reliable foundation design.

Factors Influencing Capacity

The bearing capacity of soil is significantly influenced by several factors, including soil type, depth, water table, and load eccentricity. Each of these factors plays a crucial role in determining the safe and effective design of foundations. Understanding these influences is essential for geotechnical engineers to ensure the stability and longevity of structures.

Soil type is one of the primary determinants of bearing capacity. Soils are generally classified into cohesive soils, such as clay, and cohesionless soils, such as sand. Cohesive soils exhibit significant strength due to the electrostatic forces between the clay particles, which can be quantified by the parameter of cohesion (c). The bearing capacity of cohesive soils increases with the cohesion value. On the other hand, cohesionless soils rely on the internal friction angle (ϕ) to resist loading. The density and arrangement of the soil particles influence the internal friction angle, with denser packing leading to higher bearing capacities.

The depth of the foundation (D_f) also affects the bearing capacity. As a foundation is placed deeper into the soil, the overburden pressure increases, which in turn increases the soil's density and strength. This is particularly relevant for cohesionless soils, where the depth significantly enhances the effective stress and, consequently, the bearing capacity. The Terzaghi bearing capacity equation incorporates the depth factor (N_q) to account for this increase in capacity with depth.

The presence of a water table near the foundation level can drastically reduce the bearing capacity of the soil. The saturation of soil pores leads to an increase in pore water pressure, which reduces the effective stress in the soil. For cohesionless soils, a high water table can lead to a condition known as liquefaction, where the soil behaves more like a liquid than a solid under loading. In cohesive soils, saturation reduces cohesion and can lead to a decrease in shear strength. Engineers must consider the highest possible level of the water table in their designs and apply appropriate safety factors to account for its fluctuating nature.

Load eccentricity refers to the offset of the applied load from the foundation's centroid. Eccentric loading can induce bending moments and shear forces in the foundation, leading to uneven stress distribution in the soil. This uneven distribution can cause tilting or differential settlement of the foundation, potentially leading to structural damage. The bearing capacity equations typically assume a centrally loaded, uniformly distributed load. When dealing with eccentric loads, adjustments must be made to account for the reduced effective area in contact with the soil and the non-uniform stress distribution. This is often achieved by reducing the presumed width or length of the foundation in the bearing capacity calculations to ensure a conservative design.

Incorporating these factors into the design and analysis of foundations requires a comprehensive understanding of soil mechanics and foundation engineering principles. Geotechnical site investigations, including soil borings and laboratory testing, provide the necessary data to evaluate these factors accurately. Advanced numerical modeling techniques can also simulate various loading conditions and foundation depths, incorporating the effects of the water table and load eccentricity to predict the foundation's behavior under expected loads. Through careful consideration of these factors, engineers can design foundations that ensure the safety and stability of structures under various environmental and loading conditions.

Foundation Types

Shallow Foundations Design

Designing shallow foundations, including spread footings, wall footings, and mat foundations, requires a comprehensive understanding of soil mechanics and structural loads to ensure stability and prevent excessive settlement. Spread footings are widely used to distribute the load from a single column, whereas wall footings support load-bearing walls. Mat foundations, or rafts, are employed when loads from individual columns or walls must be distributed over a larger area due to weak soil conditions or to reduce differential settlement.

Spread footings are designed based on the principle that the soil pressure distribution beneath the footing should be uniform at safe bearing capacity levels. The size of the footing is determined by dividing the total load by the allowable soil pressure, considering both dead and live loads. The depth of the footing is influenced by frost line requirements, soil type, and the presence of water tables. The footing depth must also provide adequate anchorage for reinforcement and be sufficient to resist overturning and sliding. The design process involves calculating the bearing capacity of the soil using formulas such as Terzaghi's bearing capacity equation, which considers soil cohesion, soil weight, and the depth and width of the footing. The equation is given by:

$$q_u = cN_c + \gamma D_f N_q + 0.5\gamma B N_\gamma$$

where c is the cohesion, γ is the unit weight of the soil, D_f is the depth of the foundation, B is the width of the foundation, and N_c, N_q, and N_γ are dimensionless bearing capacity factors that depend on the soil's angle of internal friction.

Wall footings, similar to spread footings, are designed to distribute the load of a wall over a larger area. The width and depth of the wall footing are determined based on the wall's load and the soil's bearing capacity. The key difference in design considerations between wall and spread footings lies in the linear distribution of load along the length of the wall, requiring calculations of soil pressure beneath the entire length of the footing. The design ensures that the footing has enough width to distribute the load within the allowable soil pressure limits and enough depth to prevent shear failure and provide adequate reinforcement coverage.

Mat foundations are utilized when the loads from the structure are so heavy or the allowable soil pressure so low that conventional footings would cover more than about half the building area. Mats are also used to minimize differential settlement on non-homogeneous soils or where there is a significant variation in loads across the structure. The design of mat foundations involves a detailed analysis of load distribution and soil-structure interaction. The thickness of the mat must be sufficient to resist bending moments and shear forces, with reinforcement provided to resist the tensile stresses generated. The analysis often requires the use of finite element methods to accurately model the complex interactions between the mat, the underlying soil, and the superimposed loads. The design process considers the modulus of subgrade reaction, which represents the soil's stiffness and is used to estimate the settlement and rotation of the mat under load.

In all cases, the design of shallow foundations requires careful consideration of soil properties, including bearing capacity, compressibility, and the presence of water, as well as the magnitude and distribution of loads from the structure. Geotechnical investigations provide the necessary data to inform these designs, including soil borings and laboratory tests to determine soil strength and compressibility characteristics. Proper design ensures that shallow foundations adequately support the structure above while preventing excessive settlement, ensuring the longevity and safety of the structure.

Deep Foundations: Piles and Drilled Shafts

Deep foundations are critical for structures requiring support from deeper, more stable soil layers to distribute loads and ensure stability. This section delves into the intricacies of piles and drilled shafts, alongside the mechanisms through which they transfer loads from the superstructure to the underlying soil or rock strata.

Piles, long slender columns made from materials such as concrete, steel, or timber, are driven into the ground or cast in situ to reach the necessary depth where the soil has adequate bearing capacity. The primary function of piles is to transfer the load through weak soil layers to a depth where the soil is capable of providing sufficient support. The load transfer mechanism of piles operates on both end bearing and skin friction. End bearing piles transfer load directly to a firm stratum at their tip, functioning similarly to a column. This is quantified by the formula $Q_u = A_b \cdot q_b$, where Q_u is the ultimate load capacity, A_b is the cross-sectional area of the pile tip, and q_b is the bearing capacity of the soil at the pile tip. Skin friction piles, on the other hand, transfer load through friction along their sides. The load capacity from skin friction can be calculated using $Q_s = A_s \cdot f_s$, where A_s is the surface area of the pile, and f_s is the frictional resistance per unit area.

Drilled shafts, also known as caissons or bored piles, are constructed by excavating a hole to the desired depth and then filling it with concrete. They are suited for heavy loads and can be used in conditions where driving piles might be impractical due to dense soil or rock layers. The load transfer in drilled shafts occurs through both end bearing at the shaft's base and skin friction along its perimeter. The capacity of a drilled shaft can be significantly higher than that of a driven pile due to the larger diameter and the potential to penetrate deeper into high-bearing capacity strata. The load-bearing capacity of a drilled shaft is given by $Q_u = Q_b + Q_s$, where Q_b is the end bearing capacity and Q_s is the skin friction capacity.

The design and selection of deep foundations require careful consideration of the geotechnical properties of the soil, the loads to be supported, and the environmental conditions of the site. Engineers must also consider the potential for settlement and lateral movement, which can affect the long-term performance and stability of the structure. Advanced testing, including static load tests and dynamic analysis, is often employed to accurately predict the behavior of deep foundations under load.

The successful implementation of deep foundations hinges on a thorough understanding of soil mechanics, structural dynamics, and construction techniques. By effectively transferring structural loads to competent soil layers or bedrock, deep foundations ensure the stability and durability of buildings, bridges, and other critical infrastructure, even in challenging ground conditions.

Consolidation and Differential Settlement

Consolidation and Settlement Analysis

Consolidation in geotechnical engineering is a process that involves the gradual reduction of water content in saturated soil under sustained load, leading to soil settlement. This phenomenon is critical in understanding how structures interact with their foundation soils over time. The consolidation process can be divided into two main phases: primary consolidation and secondary consolidation. Primary consolidation occurs as the pore water pressure in the soil decreases and the soil particles rearrange themselves more closely under the load, leading to a reduction in soil volume. This phase is largely governed by Darcy's Law, which describes the flow of fluid through porous media. The rate of primary consolidation is dependent on the soil's permeability and compressibility, as well as the thickness of the soil layer and the initial excess pore water pressure. The primary consolidation process can be quantitatively described by the Terzaghi consolidation equation:

$$U = \frac{1}{H^2} \cdot \frac{C_v \cdot t}{T_v}$$

where U is the degree of consolidation, H is the drainage path length, C_v is the coefficient of consolidation, t is the time, and T_v is the time factor. This equation helps predict the settlement of soil over time under a given load, which is crucial for the design and analysis of foundation systems.

Secondary consolidation, also known as creep, begins after the primary consolidation phase has largely completed and the excess pore water pressures have dissipated. Unlike primary consolidation, secondary consolidation is attributed to the viscous behavior of the soil matrix and the adjustment of soil particles in the microstructure over time under constant effective stress. This phase of consolidation is much slower and can continue for years or even decades. The rate of secondary consolidation is represented by the coefficient of secondary compression, C_α, which is much smaller than the coefficient of primary consolidation, C_v.

The time rate of settlement, an essential aspect of consolidation analysis, is determined by observing the change in soil volume over time. It provides valuable insights into how quickly a structure might experience settlement and allows engineers to design foundations that minimize potential damage to the structure. Various laboratory tests, such as the oedometer test (one-dimensional consolidation test), are used to measure the consolidation characteristics of soils. These tests involve applying a series of loads to a soil sample confined laterally in a rigid ring and measuring the amount and rate of settlement that occurs. The data obtained from these tests are used to plot void ratio (e) versus effective stress (σ') curves and to calculate the coefficients of consolidation and secondary compression.

Understanding the consolidation behavior of soils is paramount for the design of stable and durable structures. It enables engineers to predict and mitigate potential settlement issues, ensuring the longevity and safety of buildings, bridges, and other infrastructure. Through meticulous analysis and testing, geotechnical engineers can develop effective foundation solutions that account for the complex interactions between soil and structure, ultimately contributing to the successful completion of engineering projects.

Settlement Effects and Mitigation Techniques

Differential settlement occurs when a structure's foundation settles unevenly, potentially leading to significant structural damage and, in severe cases, failure. This phenomenon is primarily caused by variations in soil properties, load distribution, and environmental factors. Understanding the causes and implementing effective mitigation techniques is crucial for ensuring the longevity and safety of structures.

Variations in soil properties, such as strength, compressibility, and moisture content, can lead to differential settlement. Soils with heterogeneous layers, where stiff soil strata overlay compressible layers, are particularly susceptible. The differential loading, where parts of a structure impose different pressures on the foundation soil, exacerbates this issue. Environmental factors, including erosion, soil moisture changes due to vegetation or drainage alterations, and seismic activity, can also contribute to uneven settlement.

Mitigation of differential settlement begins with a comprehensive geotechnical investigation to understand the soil profile and properties at the construction site. Soil testing, including standard penetration tests (SPT), cone penetration tests (CPT), and oedometer tests, provides valuable data on soil behavior under load. Based on these findings, engineers can design foundations that accommodate or correct potential differential settlement issues.

One common mitigation technique is the use of deep foundations, such as piles or drilled shafts, which transfer loads to deeper, more stable soil layers or bedrock. This approach is particularly effective in areas with significant soil variability or where surface soils have low bearing capacity. The design of deep foundations requires careful consideration of load transfer mechanisms, including end bearing and skin friction, to ensure that the structure is supported adequately.

Another technique involves soil improvement methods to enhance the bearing capacity and reduce the compressibility of the foundation soil. Techniques such as compaction, grouting, and the use of geosynthetics can stabilize soil and make it more uniform, thereby reducing the risk of differential settlement. For example, controlled modulus columns (CMCs) or stone columns can be installed to reinforce soft soils, while jet grouting can be used to create more uniform soil conditions by mixing the in-situ soil with a cementitious grout.

Preloading is a method used to consolidate soft soils and reduce future settlement under structural loads. By applying a temporary load to the soil before construction, excess pore water pressure is dissipated, and the soil is compacted. This process accelerates consolidation, reducing the amount of settlement that occurs after the structure is built. Vertical drains, such as wick drains, can be installed to expedite the consolidation process by providing a path for water to escape from the soil.

In cases where differential settlement cannot be entirely prevented, structural solutions may be employed to accommodate movement without compromising the integrity of the structure. These solutions include the use of flexible connections between structural elements, adjustable foundations that can be leveled after construction, and the strategic placement of expansion joints to allow for movement.

The successful mitigation of differential settlement requires a multidisciplinary approach that combines geotechnical investigation, innovative foundation design, soil improvement techniques, and, when necessary, structural adaptations. By addressing the causes of differential settlement and implementing targeted mitigation strategies, engineers can ensure that structures remain safe, functional, and durable over their intended lifespan.

Slope Stability

Slope Failure Mechanisms

Understanding slope failure mechanisms requires a deep dive into the types of failures and the conditions that precipitate such events. Slope failures are primarily categorized based on the material involved (soil or rock), the type of movement, and the moisture content. The common types of slope failures include **rotational slides**, **translational slides**, **flows**, and **falls**. Each of these failure mechanisms has distinct characteristics and triggers.

Rotational slides occur along a curved surface of rupture. This type of failure is typical in homogeneous, fine-grained soils where the base of the slide moves outward and downward, causing the upper portion to rotate about an axis parallel to the slope. The failure surface is concave upwards, resembling a spoon-shaped curve. The mathematical representation of the failure surface can be approximated by the equation of a circle, where the center of rotation is located below the original ground surface. The stability of a slope against rotational failure can be analyzed using the **Swedish Circle Method** or **Bishop's Simplified Method**, which consider the balance of moments about the center of rotation.

Translational slides involve the movement of a soil or rock mass along a planar or gently undulating surface. These slides are common in slopes with distinct stratification, where a weaker layer underlies a stronger, overlying stratum. The failure plane is often parallel to the slope surface, leading to a block of material sliding down the slope. The stability of slopes against translational sliding can be assessed using the **Limit Equilibrium Method (LEM)**, which evaluates the factor of safety against sliding by comparing the driving forces to the resisting forces along the potential failure plane.

Flows are a type of slope failure characterized by the material behaving fluidly. This can occur in both soil and rock, depending on the water content. Soil flows, such as **earthflows** and **mudflows**, are common in saturated soils where the pore water pressure reduces the effective stress, thereby decreasing the soil's shear strength. Rock flows, or **rock avalanches**, involve the rapid movement of fragmented rock down a slope. The analysis of flow stability often requires understanding the rheological properties of the material and may involve complex computational models to predict the flow behavior and potential runout distance.

Falls involve the free fall, bouncing, or rolling of individual rocks or soil clumps from steep slopes or cliffs. This mechanism is more common in rock slopes or heavily fractured soils. The stability against falls is often assessed through rockfall trajectory modeling, which considers the geometry of the slope, the properties of the falling mass, and the energy dissipation during impact with the slope surface.

The stability of slopes and the likelihood of failure are influenced by several factors, including the **geotechnical properties** of the slope material, **slope geometry**, **water content**, and **external**

loads or changes. The **geotechnical properties**, such as cohesion, internal friction angle, and unit weight, are fundamental in determining the shear strength of the material. **Slope geometry**, including slope angle and height, directly impacts the gravitational forces acting on the slope material. **Water content** plays a critical role in slope stability, as it affects the pore water pressure and, consequently, the effective stress and shear strength of the soil. **External loads**, such as construction activities, excavation, or added weight on the slope crest, can also precipitate failure by altering the stress distribution within the slope.

The analysis and mitigation of slope failures require a comprehensive understanding of the failure mechanisms, the conditions under which they occur, and the factors that influence slope stability. By applying appropriate analytical methods and considering the specific characteristics of each slope, engineers can design effective stabilization measures to prevent slope failures and protect infrastructure and lives.

Stability Analysis for Embankments, Cuts, and Dams

The stability analysis of embankments, cuts, and dams using **limit equilibrium methods (LEM)** is a critical aspect of geotechnical engineering, aimed at preventing failure and ensuring the safety and longevity of these structures. The LEM is predicated on the assumption that a slope or structure is on the verge of failure along a potential slip surface. The method involves dividing the mass into slices and calculating the forces acting on each slice to determine if the structure is stable or not. The factor of safety (F_s) is a key outcome of this analysis, representing the ratio of resisting forces to driving forces. A F_s greater than 1 indicates stability, while a value less than 1 suggests potential failure.

For **embankments**, the stability analysis focuses on the potential for sliding due to the weight of the embankment material and any applied loads, including vehicles or structures. The analysis must consider the embankment geometry, soil properties, pore water pressures, and the presence of any reinforcing materials. The **Bishop Simplified Method** is commonly used for circular failure surfaces, given by:

$$F_s = \frac{\sum(c' \cdot L + (W \cdot \sin\theta) \cdot \tan\phi')}{\sum(W \cdot \cos\theta)}$$

where c' is the effective cohesion, L is the length of the arc slice base, W is the weight of the slice, θ is the inclination angle of the slice base, and ϕ' is the effective angle of internal friction.

For **cuts**, or excavations, stability analysis must address the risk of slope failure, which can endanger both construction operations and adjacent structures. The analysis considers the cut geometry, soil stratification, groundwater conditions, and the temporary or permanent support systems in place. The **Janbu Simplified Method** is often applied for non-circular failure surfaces, focusing on the balance of moments and normal forces to evaluate stability.

Dams present a complex challenge for stability analysis due to their size, the variability of materials used in construction, and the significant hydrostatic pressures exerted by stored water. The stability analysis of dams typically involves assessing both the upstream (wet) and downstream (dry) slopes, considering the potential for sliding and overturning. The **Morgenstern-Price Method** is frequently employed for more complex geometries and loading conditions, incorporating both force and moment equilibrium in a rigorous manner:

$$F_s = \frac{\sum(c' \cdot L + W \cdot \sin\theta \cdot \tan\phi' + U \cdot \tan\phi')}{\sum(W \cdot \sin\theta + U)}$$

where U represents the hydrostatic uplift forces acting on the potential sliding mass.

The selection of the appropriate limit equilibrium method depends on the specific conditions of the project, including the geometry of the structure, the type of soil or rock, the presence of water, and the complexity of the potential failure mechanism. Advanced software tools are often used to perform these analyses, allowing for the consideration of multiple failure surfaces and the integration of complex soil models.

The accurate determination of soil and rock properties is crucial for reliable stability analysis. This includes the cohesion (c'), angle of internal friction (ϕ'), unit weight (γ), and the assessment of pore water pressures (u). Field investigations and laboratory testing provide the necessary data to inform these parameters.

The application of limit equilibrium methods in the stability analysis of embankments, cuts, and dams is a testament to the importance of understanding and mitigating the risks associated with slope instability. Through meticulous analysis and the judicious use of engineering judgment, geotechnical engineers play a vital role in safeguarding these critical structures against failure, thereby protecting both lives and investments.

Soil Stabilization

Chemical Stabilization Additives

Continuing from the foundational understanding of soil stabilization techniques, chemical stabilization involves the addition of specific additives such as **lime**, **cement**, and **fly ash** to enhance the physical and mechanical properties of soil. These materials interact with soil particles to alter their characteristics, improving strength, reducing plasticity, and enhancing load-bearing capacity.

Lime (calcium hydroxide) is particularly effective in treating clayey soils. The addition of lime initiates a series of chemical reactions known as **pozzolanic reactions**, where lime chemically reacts with silica and alumina present in the clay particles, forming cementitious compounds over time. This reaction not only reduces the soil's plasticity but also increases its strength. The effectiveness of lime stabilization can be quantified by evaluating the change in the **Plasticity Index (PI)** and the **Unconfined Compressive Strength (UCS)** of the treated soil. The typical equation governing the pozzolanic reaction is represented as:

$$Ca(OH)_2 + SiO_2 + H_2O \rightarrow CaSiO_3 \cdot H_2O + Heat$$

This reaction, which produces calcium silicate hydrate, is exothermic and contributes to the long-term gain in strength.

Cement, another widely used stabilizer, introduces calcium silicates and aluminates that also engage in pozzolanic reactions with the soil. The hydration of cement particles results in the formation of a hardened matrix that binds soil particles together, significantly improving the soil's load-bearing capacity. The general reaction for cement hydration can be simplified as:

$$Ca_3SiO_5 + H_2O \rightarrow CaSiO_3 \cdot H_2O + Ca(OH)_2$$

This process leads to the development of a strong and durable soil-cement matrix, enhancing the soil's structural integrity.

Fly ash, a byproduct of coal combustion in power plants, contains silica, alumina, and calcium oxide, which can react with water and soil to form cementitious compounds similar to those produced by lime and cement. The presence of **Class C fly ash** is particularly beneficial for soil stabilization due to its self-cementing properties. Fly ash particles fill the voids between soil grains, reducing the water content and increasing the density and strength of the soil. The reaction involving fly ash can be represented as follows:

$$CaO + SiO_2 + H_2O \rightarrow CaSiO_3 \cdot H_2O$$

This reaction, similar to the pozzolanic reaction, contributes to the soil's improved mechanical properties.

The selection of an appropriate stabilizing agent (lime, cement, or fly ash) depends on the specific soil properties, desired outcomes, and environmental considerations. Each additive has its unique advantages and limitations, and their effectiveness can be maximized by tailoring the treatment to the specific characteristics of the soil in question. The application of these chemical stabilizers transforms the soil into a more stable and durable construction material, capable of supporting various types of infrastructure projects.

Geosynthetics: Geotextiles and Geogrids

Geosynthetics, including geotextiles and geogrids, play a pivotal role in modern soil stabilization techniques, offering innovative solutions for reinforcement and drainage that significantly enhance the performance and durability of geotechnical engineering projects. Geotextiles, made from synthetic fibers, can be woven or non-woven and are utilized extensively for separation, filtration, protection, and drainage in a variety of applications. Their primary function is to prevent the intermixing of different soil layers while allowing water to pass through, thereby maintaining structural integrity and preventing erosion or subsidence. The permeability of geotextiles is a critical property, determined by the fabric's structure and the size of its pores, which must be selected based on the specific requirements of the project, including soil particle size and expected water flow rates.

Geogrids, on the other hand, are characterized by their open grid-like structure, which provides superior reinforcement to soil and other materials. Made from polymers such as polyethylene or polypropylene, geogrids are designed to interlock with surrounding materials, enhancing load distribution and contributing to the stability of slopes, retaining walls, and roadbeds. The tensile strength of geogrids is leveraged to counteract forces that could lead to soil deformation or failure, making them an essential component in the design and construction of embankments, foundations, and pavement systems.

The integration of geosynthetics into soil stabilization projects requires careful consideration of material properties, environmental conditions, and project objectives. For instance, the selection between woven and non-woven geotextiles will depend on the need for higher filtration rates or greater tensile strength, respectively. Similarly, the choice of geogrid type, whether uniaxial or biaxial, will be influenced by the direction of the expected loads and the nature of the soil being reinforced.

Installation techniques also play a crucial role in the effectiveness of geosynthetics. Proper overlap, anchoring, and tensioning of geotextiles and geogrids are critical to ensure that they perform as intended, without failure or displacement over time. Additionally, the compatibility of geosynthetics with the surrounding environment, including resistance to biological and chemical degradation, must be assessed to ensure long-term performance.

The benefits of using geosynthetics in soil stabilization are manifold. They include the reduction of construction time and costs, increased lifespan of structures, and the minimization of maintenance requirements. Moreover, geosynthetics offer environmentally friendly solutions by reducing the need for natural aggregate materials and enabling the use of local soils that might otherwise be considered unsuitable for construction.

The application of geotextiles and geogrids in soil stabilization represents a convergence of engineering innovation and practical efficiency. By understanding the properties and functions of these materials, engineers can design and implement solutions that address the complex challenges of modern geotechnical engineering projects. Through the strategic use of geosynthetics, it is possible to achieve enhanced performance, sustainability, and resilience in infrastructure development, making them an indispensable tool in the field of soil stabilization.

Chapter 13: Transportation Engineering

Geometric Design

Roadway Design Basics

Design speed serves as a fundamental parameter in the geometric design of roadways, influencing the selection of horizontal and vertical alignments, cross-sectional elements, and safety features. It is defined as the maximum safe speed that can be maintained over a specified section of road under favorable weather conditions and daylight, assuming a typical vehicle and an average driver. The American Association of State Highway and Transportation Officials (AASHTO) provides guidelines for determining design speeds based on the functional classification of the roadway, anticipated traffic volumes, and surrounding land use.

Horizontal alignment encompasses the plan view of the road, detailing the straight paths (tangents) and curves. The design of horizontal curves is directly influenced by the selected design speed, as higher speeds require larger radii to ensure vehicle stability and comfort. The minimum radius R of a curve for a given design speed V (in mph) and a coefficient of friction f can be calculated using the formula:

$$R = \frac{V^2}{15(f + e)}$$

where e is the superelevation rate, or the banking of the roadway, expressed as a decimal. Superelevation counteracts the lateral acceleration experienced by a vehicle on a curve, enhancing safety and comfort.

Vertical alignment refers to the longitudinal profile of the roadway, consisting of grades (slopes) and vertical curves. Grades are selected based on the design speed, type of vehicles using the road, and environmental considerations such as drainage and visibility. Vertical curves provide a transition between different grades, ensuring that the change in slope is gradual. The length of a vertical curve L is determined based on the design speed V, the algebraic difference between the grades A, and a comfort parameter K, which varies with the type of vertical curve (sag or crest):

$$L = KV$$

where K is selected from standard tables provided by AASHTO, ensuring that drivers experience acceptable levels of comfort and visibility.

The cross-section of a roadway includes the travel lanes, shoulders, medians, and drainage facilities. The width of travel lanes is influenced by the design speed, with higher speeds necessitating wider lanes to accommodate vehicle dynamics and driver reaction times. Shoulders provide a recovery area for errant vehicles and space for emergency stops and maintenance activities; their width is also correlated with the design speed and the volume of traffic. Medians separate opposing flows of traffic and contribute to safety, especially on high-speed, multi-lane highways. The design of drainage facilities, including ditches and culverts, ensures the efficient removal of surface water, preventing hydroplaning and maintaining the structural integrity of the roadway.

Incorporating these elements into the geometric design of roadways requires a comprehensive understanding of vehicle dynamics, driver behavior, and environmental factors. Engineers must balance safety, efficiency, and cost considerations, applying principles of physics and mathematics to create roadways that facilitate smooth and safe travel at the intended design speeds. The application of these design principles ensures that roadways are capable of accommodating current and future traffic demands, providing a durable infrastructure that supports economic growth and community development.

Intersection Design

Roundabouts, as an integral component of intersection design, offer a sustainable and efficient traffic management solution. Unlike traditional signalized intersections, roundabouts reduce the severity of crashes by eliminating the potential for right-angle and head-on collisions. The design of a roundabout requires careful consideration of geometric features, including entry and circulatory lane widths, central island diameter, and approach angles, to accommodate various vehicle sizes, including emergency and freight vehicles. The entry curvature is designed to slow vehicles to a safe speed as they enter the roundabout, typically between 15 to 25 mph, enhancing safety and reducing congestion. The inscribed circle diameter (ICD) is a critical dimension, influencing the roundabout's capacity and the ease of navigation for large vehicles. The ICD is determined based on the expected traffic volume and the mix of vehicle types, with larger diameters providing greater capacity and accommodation.

Signalized intersections, on the other hand, rely on traffic signals to control vehicle and pedestrian flows, requiring a different set of design considerations. The timing of signals is crucial, with phases designed to optimize traffic flow and minimize delays. This involves the calculation of cycle lengths, green splits, and pedestrian crossing times, tailored to the intersection's specific traffic patterns. Advanced traffic signal systems may incorporate adaptive signal control technology, which adjusts signal timings in real-time based on traffic conditions, further improving efficiency and reducing congestion.

Sight distance is a paramount safety consideration in the design of both roundabouts and signalized intersections. Adequate sight distance allows drivers to see other vehicles, pedestrians, and obstacles in time to react safely. At roundabouts, the sight distance must be sufficient for a driver to detect an approaching vehicle from the left and decide whether to enter the roundabout. This is typically calculated based on the approach speed, with longer distances required for higher speeds. The American Association of State Highway and Transportation Officials (AASHTO) provides guidelines for minimum sight distances at intersections, including stopping sight distance (SSD) and decision sight distance (DSD). SSD is the distance a driver needs to stop after seeing an obstacle in their path, while DSD is the distance needed to make an informed decision, such as turning or adjusting speed, safely.

For signalized intersections, sight distance considerations also include ensuring that drivers have an unobstructed view of traffic signals and that pedestrian crosswalks are visible from a safe stopping distance. Additionally, the placement of physical elements such as signage, landscaping, and street furniture must not impede sight lines between drivers, pedestrians, and cyclists.

The design of intersections, whether roundabouts or signalized, requires a holistic approach that balances traffic efficiency, safety, and environmental considerations. Engineers must apply rigorous analysis and adhere to established guidelines to create intersections that facilitate smooth and safe movement for all users. Through careful planning and design, intersections can significantly contribute to the overall performance of the transportation network, supporting economic activity and enhancing the quality of life in communities.

Pavement System Design

Pavement Structure Essentials

Layer thickness in pavement design is a critical factor that directly influences the durability and performance of the roadway. The determination of appropriate layer thicknesses involves understanding the interplay between traffic loads, environmental conditions, and the material properties of each layer. The American Association of State Highway and Transportation Officials (AASHTO) provides guidelines for calculating layer thickness based on the expected Equivalent Single Axle Loads (ESALs) over the pavement's design life. The structural number (SN), a dimensionless quantity, is used in these calculations to represent the overall strength of the pavement structure, which is a function of the thickness and material properties of the layers. The general formula for determining the required structural number is given by:

$$SN = a_1 D_1 + a_2 D_2 m_2 + a_3 D_3 m_3$$

where D_1, D_2, and D_3 are the thicknesses of the surface, base, and subbase layers, respectively, a_1, a_2, and a_3 are the layer coefficients that represent the strength of each material, and m_2 and m_3 are the drainage coefficients for the base and subbase layers, respectively.

Subgrade preparation is another fundamental aspect of pavement structure that affects the longevity and effectiveness of the pavement system. The subgrade is the natural soil upon which the pavement structure is constructed, and its preparation involves grading, compaction, and stabilization processes to ensure a stable foundation for the overlying layers. Proper subgrade preparation aims to achieve a uniform support capability, minimize differential settlements, and provide adequate drainage. The California Bearing Ratio (CBR) test is commonly used to assess the load-bearing capacity of the subgrade soil, and the results are used to design the thickness of the pavement layers.

Material selection for each layer of the pavement structure is based on the desired performance characteristics, cost considerations, and availability of materials. Asphalt concrete and Portland cement concrete are commonly used for surface layers due to their ability to withstand traffic loads and environmental conditions. The base and subbase layers typically consist of granular materials, such as crushed stone or gravel, which are selected for their drainage properties and mechanical stability. Recycled materials, such as reclaimed asphalt pavement (RAP) and recycled concrete aggregate (RCA), are increasingly being used in pavement layers to promote sustainability and reduce costs.

Drainage considerations are integral to pavement design, as water infiltration is a primary cause of pavement deterioration. Effective drainage systems prevent the accumulation of water within or beneath the pavement structure, thereby protecting the pavement from the detrimental effects of water-induced damage, such as stripping, rutting, and frost heave. The design of pavement drainage involves the installation of surface drainage features, such as curbs and gutters, and subsurface drainage systems, including perforated pipes and aggregate drains. The permeability of the base and subbase materials is a critical factor in the effectiveness of the drainage system, and proper material selection and compaction practices are essential to ensure adequate drainage capacity.

The integration of these elements—layer thickness, subgrade preparation, material selection, and drainage considerations—into the pavement design process is essential for the construction of durable and efficient roadways. By adhering to established guidelines and employing sound engineering judgment, transportation engineers can design pavement structures that meet the demands of traffic loads and environmental conditions, ensuring the safety and mobility of the traveling public.

Rehabilitation Techniques

Overlay design is a critical rehabilitation technique for extending the life of existing pavements by adding a new layer of asphalt or concrete. The primary objective is to improve structural capacity and smoothness while minimizing the impact on the existing pavement. The design process involves determining the appropriate overlay thickness, which is influenced by factors such as the existing pavement condition, traffic loads, and desired service life. The American

Association of State Highway and Transportation Officials (AASHTO) Guide for Design of Pavement Structures provides methodologies for calculating overlay thickness based on empirical data and mechanistic-empirical principles. For asphalt overlays, the thickness typically ranges from 1.5 to 4 inches, depending on the existing pavement's condition and the overlay material's properties.

Resurfacing is another pivotal rehabilitation strategy, focusing on removing part of the existing pavement surface and replacing it with a new layer. This technique is often employed to address surface distresses such as raveling, cracking, and rutting, which can significantly affect pavement performance and user safety. Resurfacing not only restores the structural integrity and ride quality of the pavement but also provides an opportunity to improve surface characteristics such as skid resistance and reflectivity. The decision to resurface and the selection of materials are based on a comprehensive evaluation of the pavement's current condition, traffic characteristics, and environmental factors.

Pavement maintenance strategies are essential for preserving the functional and structural aspects of the pavement, thereby extending its service life and optimizing the investment in infrastructure. Maintenance activities can be categorized into preventive, corrective, and emergency actions. Preventive maintenance includes applications such as crack sealing, seal coating, and thin overlays designed to protect the pavement from environmental damage and delay deterioration. Corrective maintenance addresses existing defects that affect pavement performance, such as patching potholes, repairing cracks, and fixing drainage issues. Emergency maintenance is conducted in response to unforeseen events like natural disasters or accidents that cause immediate harm to the pavement structure.

The implementation of a well-planned maintenance program, guided by principles such as timely intervention and the use of appropriate materials and techniques, is crucial for sustaining pavement performance. Regular monitoring and evaluation of pavement condition through visual inspections and performance measurements support the identification of maintenance needs and the prioritization of rehabilitation efforts. By adopting a proactive approach to pavement maintenance, transportation agencies can ensure the safety, functionality, and longevity of road networks, ultimately contributing to efficient and reliable transportation systems.

Traffic Capacity and Flow Theory

Flow Fundamentals in Traffic Analysis

Understanding the fundamental principles of traffic flow is crucial for effective traffic analysis and management. The three primary characteristics of traffic flow are **speed** (v), **density** (ρ), and **flow** (q). These elements are interrelated and form the basis for analyzing and designing traffic systems.

Speed (v) refers to the distance covered by a vehicle per unit of time. It is typically measured in miles per hour (mph) or kilometers per hour (km/h). Speed is a critical factor in traffic flow analysis as it influences the capacity and safety of roadways.

Density (ρ) is defined as the number of vehicles per unit length of the roadway, usually expressed in vehicles per mile (vpm) or vehicles per kilometer (vpk). High density indicates congested conditions, while low density suggests underutilized roadways.

Flow (q), or traffic volume, represents the rate at which vehicles pass a reference point on the roadway, typically measured in vehicles per hour (vph). The relationship between flow, speed, and density is given by the fundamental traffic flow equation: $q = \rho \cdot v$.

This equation highlights the direct proportionality between flow and speed at a constant density, and between flow and density at a constant speed. However, in real-world conditions, speed and density are inversely related; as density increases, the speed typically decreases due to the closer proximity of vehicles, leading to a nonlinear relationship between flow and density.

The analysis of these relationships is essential for understanding the performance of traffic systems under various conditions. For instance, at low densities, vehicles can travel at higher speeds, resulting in a linear increase in flow. As density continues to increase, interactions among vehicles become more frequent, causing speeds to decrease and flow to grow at a diminishing rate. Eventually, a point of maximum flow is reached, beyond which further increases in density lead to a decrease in flow due to congestion.

Graphical representations, such as the **speed-density curve** and the **flow-density curve**, are instrumental in visualizing these relationships. The speed-density curve illustrates the inverse relationship between speed and density, while the flow-density curve, often bell-shaped, shows the initial increase in flow with density, reaching a peak, and then decreasing as congestion sets in.

In traffic analysis, understanding these fundamental principles allows engineers to assess roadway performance, design efficient traffic systems, and implement measures to mitigate congestion. For example, by analyzing flow-density curves, engineers can determine the capacity of a roadway and identify the densities at which traffic flow becomes unstable. This information is critical for planning roadway expansions, setting speed limits, and designing traffic control systems that optimize flow and enhance safety.

Moreover, the application of these principles extends to the design of intelligent transportation systems (ITS) that use real-time data to manage traffic flow dynamically. Adaptive signal control, variable speed limits, and congestion pricing are examples of strategies informed by traffic flow theory to improve roadway efficiency and safety.

The interplay between speed, density, and flow forms the foundation of traffic capacity and flow theory. By applying these concepts, transportation engineers can design and manage roadways that accommodate varying traffic demands, minimize congestion, and promote safe and efficient travel.

Capacity Analysis Using HCM Principles

Capacity analysis, a critical component of traffic engineering, employs the Highway Capacity Manual (HCM) principles to evaluate the performance of highways, intersections, and ramps under various traffic conditions. The HCM provides methodologies for quantifying roadway capacity, service levels, and performance measures, enabling engineers to design and manage traffic systems effectively.

Highway Capacity Analysis focuses on determining the maximum number of vehicles that can traverse a segment of a highway under prevailing conditions. The HCM introduces the concept of **Level of Service (LOS)**, a qualitative measure describing operational conditions within a traffic stream, characterized by factors such as speed, travel time, freedom to maneuver, traffic interruptions, and comfort and convenience. LOS ranges from A (free flow) to F (forced or breakdown flow), with each level providing insights into the efficiency and quality of the traffic flow. The fundamental formula for highway capacity is given by $C = \dfrac{1000}{H}$, where C represents the capacity in vehicles per hour per lane, and H is the average headway in seconds between vehicles. This formula underscores the inverse relationship between vehicle headway and traffic flow capacity.

Intersection Capacity Analysis evaluates the ability of intersections to accommodate various traffic volumes without undue delays. The HCM delineates methodologies for analyzing the capacity of signalized and unsignalized intersections, incorporating parameters such as traffic volumes, signal timing, phasing, and geometric design. For signalized intersections, the critical measure is the **saturation flow rate**, the maximum hourly rate at which vehicles can be served per lane during a green signal phase, typically expressed in vehicles per hour green (vphg). The saturation flow rate can be adjusted for factors like lane width, grade, and turning movements. The capacity of an intersection is influenced by the signal timing, with longer green phases for a particular movement increasing that movement's capacity.

Ramp Capacity Analysis involves assessing the capability of on-ramps and off-ramps to handle traffic volumes efficiently, a key factor in preventing bottlenecks on highways. Ramp capacity is significantly affected by the merge and diverge characteristics, including the geometry of the ramp, the presence of auxiliary lanes, and the traffic control devices in place. The HCM provides guidelines for estimating ramp capacity by considering factors such as ramp length, lane width, and the volume of merging or diverging traffic. The analysis of ramp capacity is crucial for the design and operation of freeway systems, ensuring smooth transitions between freeways and arterial roads.

The application of HCM principles in capacity analysis requires a comprehensive understanding of traffic flow theories and the ability to interpret and apply empirical data. Traffic engineers must consider a wide range of variables, including vehicle characteristics, driver behavior, roadway conditions, and environmental factors. By employing HCM methodologies, engineers can identify potential bottlenecks, evaluate the impact of proposed roadway improvements, and develop strategies to enhance traffic flow and safety.

The HCM's approach to capacity analysis emphasizes the dynamic nature of traffic engineering, encouraging the use of simulation models and real-time data to adapt to changing traffic patterns. This adaptability is essential for addressing the challenges of modern transportation systems, where fluctuations in traffic demand and the integration of intelligent transportation systems necessitate flexible and responsive traffic management strategies.

Through the meticulous application of HCM principles, transportation engineers can ensure that highways, intersections, and ramps are designed and operated to meet the needs of all users, promoting efficient movement, reducing congestion, and improving overall roadway safety.

Traffic Control Devices

Traffic Signals and Signs

The Manual on Uniform Traffic Control Devices (MUTCD) sets forth the standards used by road managers nationwide to install and maintain traffic control devices on all public streets, highways, bikeways, and private roads open to public travel. The MUTCD is published by the Federal Highway Administration (FHWA) and details the guidelines for the placement, design, and use of traffic signs, signals, and road markings. Understanding the MUTCD standards is crucial for engineers to design and implement traffic control systems that ensure road safety and efficiency.

Traffic signals are critical components of the transportation infrastructure, designed to control vehicle and pedestrian traffic flow. The timing of traffic signals is a complex process that involves the analysis of traffic volume, speed, and density, along with pedestrian crossing needs and adjacent signal timings. Signal timing strategies include fixed time, actuated, and adaptive signal control. Fixed-time signal control operates on a preset timing schedule, which does not change in response to traffic conditions. Actuated signal control adjusts the duration of green lights based on the detection of vehicles or pedestrians, typically through inductive loops or cameras. Adaptive signal control systems dynamically adjust signal timings in real-time based on current traffic conditions using advanced algorithms and traffic detection systems.

The placement of traffic signs must adhere to the MUTCD standards, which specify the height, size, reflectivity, and location for various types of signs to ensure they are visible and understandable to drivers and pedestrians. Signs must be placed at specific distances from intersections, curves, and other roadway features to provide adequate reaction time for road users. The MUTCD also categorizes signs into regulatory, warning, and guide signs, each serving a specific purpose in traffic control and road safety.

Regulatory signs inform road users of traffic laws and regulations and must be obeyed. Examples include stop signs, yield signs, speed limit signs, and no parking signs. Warning signs alert drivers to potential hazards or changes in road conditions, such as curves, intersections, pedestrian crossings, and school zones. Guide signs provide directional and mileage information to help drivers navigate to destinations.

The MUTCD standards ensure uniformity and consistency in traffic control devices across the United States, which is essential for the safety and efficiency of the transportation system. Engineers and transportation professionals must have a thorough understanding of these standards to design and implement effective traffic control solutions. Compliance with the MUTCD enhances road safety, reduces the likelihood of traffic accidents, and improves the overall flow of traffic. By adhering to these guidelines, engineers contribute to creating a safer and more predictable driving environment, which is critical for the well-being of all road users.

Road Markings and Intelligent Traffic Systems

Pavement markings play a crucial role in guiding and informing road users, enhancing the safety and efficiency of the transportation network. These markings include lines, arrows, symbols, and words painted or inlaid onto the road surface to delineate traffic lanes, indicate traffic flow directions, highlight critical safety zones, and convey regulations directly applicable to specific lanes or movements. The effectiveness of pavement markings depends on their visibility, both in daylight and at night under vehicle headlights, which is achieved through the use of reflective materials such as glass beads embedded in the paint or thermoplastic materials that provide high retroreflectivity.

The design and application of pavement markings must comply with the standards set forth in the Manual on Uniform Traffic Control Devices (MUTCD), which specifies the types, dimensions, colors, and locations of markings to ensure consistency and recognizability across the United States. For instance, white lines are used to separate lanes of traffic moving in the same direction and to mark the right edge of the roadway, while yellow lines separate lanes of traffic moving in opposite directions and mark the left edge of roadways on divided highways. Stop lines, crosswalks, and symbols such as turn arrows are also defined by specific criteria in the MUTCD to optimize their visibility and comprehension by drivers and pedestrians alike.

In addition to traditional pavement markings, the integration of advanced systems and technologies is transforming the landscape of traffic control and management. Intelligent Traffic Systems (ITS) encompass a broad range of applications that monitor, manage, and enhance the flow of traffic. These systems utilize sensors, cameras, and other detection devices to collect real-time data on traffic conditions, which is then processed and used to dynamically control traffic signals, manage congestion, provide traveler information, and improve the overall safety and efficiency of the transportation network.

Dynamic signage, a key component of ITS, includes variable message signs (VMS) and dynamic speed limit signs that can change their displays based on current traffic conditions, incidents, weather, or roadwork. VMS can provide real-time information to drivers about congestion, accidents, travel times, and alternative routes, enabling them to make informed decisions and adjust their driving accordingly. Dynamic speed limit signs can adjust speed limits in response to varying traffic conditions, weather, and road surface conditions, helping to reduce the risk of accidents and improve traffic flow.

The integration of pavement markings with ITS and dynamic signage represents a holistic approach to traffic control that leverages both traditional and advanced technologies to enhance road safety and efficiency. For example, smart pavement markings that are visible under specific weather conditions or that can change color to warn drivers of upcoming hazards are currently under development. These innovations, combined with ITS, promise to significantly improve the way traffic is managed and controlled, leading to safer, more efficient, and more sustainable transportation systems.

The implementation of advanced traffic control systems requires a multidisciplinary approach that encompasses traffic engineering, information technology, and urban planning. Engineers must ensure that these systems are designed and deployed in a manner that complements existing traffic control devices and meets the needs of all road users, including motorists, cyclists, and pedestrians. This involves careful planning, design, testing, and evaluation to ensure that the benefits of these technologies are fully realized while minimizing potential disruptions to the transportation network.

As the transportation infrastructure continues to evolve, the role of pavement markings and advanced systems in traffic control will undoubtedly expand, driven by advancements in technology and the growing need for more intelligent and responsive traffic management solutions. The successful integration of these elements will be critical in addressing the challenges of modern transportation, including congestion, safety, and environmental impact, and in creating a more efficient and sustainable future for road travel.

Transportation Planning

Travel Demand Modeling: Trip Generation & Distribution

Travel demand modeling is a critical component in the planning and operation of transportation systems, providing insights into how people move within an urban environment. It encompasses several key concepts: trip generation, trip distribution, and modal split, each of which plays a vital role in understanding and forecasting travel behavior.

Trip Generation is the first step in the travel demand modeling process. It predicts the number of trips originating from or destined to a particular area within a given time period. The primary focus is on identifying the characteristics of land use, such as residential, commercial, or industrial areas, and correlating these with the number of trips generated. For instance, a residential area's trip generation rate might be determined by factors such as household size, income level, and car ownership rates. Mathematical models, often regression-based, are used to estimate the relationship between these factors and the number of trips produced or attracted by different land uses.

Trip Distribution extends the analysis by determining the destination of the trips generated in the first step. It involves modeling the spatial interaction between different zones within a study area, taking into account the distance or travel time between zones, the attractiveness of the destination zones, and the availability of transportation facilities. Gravity models are commonly used for this purpose, which assume that the number of trips between two zones is directly proportional to the product of their activities (e.g., population, employment) and inversely proportional to some function of the distance or travel cost between them.

Modal Split, also known as mode choice analysis, is concerned with the selection of transportation modes (e.g., car, public transit, walking, cycling) for the trips identified in the trip generation and distribution steps. This phase of modeling assesses the factors influencing travelers' choice of mode, including travel time, cost, convenience, and personal preferences. Discrete choice models, such as logit models, are frequently employed to predict the probability of choosing a particular mode based on these factors. The outcome of the modal split analysis is crucial for estimating the demand for different modes of transportation and for planning the necessary infrastructure and services to accommodate this demand.

Incorporating these components into travel demand modeling allows urban planners and engineers to simulate current and future travel patterns, assess the impact of new developments or changes in land use, and make informed decisions about transportation policies, infrastructure investments, and service improvements. By understanding the complex interplay between land use, transportation options, and traveler behavior, stakeholders can design more efficient, sustainable, and user-friendly transportation systems that meet the needs of diverse communities.

Safety and Sustainable Transportation Strategies

Safety and sustainability in transportation engineering are paramount to creating efficient and resilient infrastructure systems that protect the environment and ensure public safety. The integration of crash analysis and multimodal planning into transportation projects is essential for identifying risk factors and implementing strategies that reduce accidents and enhance the overall sustainability of transportation networks.

Crash analysis involves the systematic examination of collision data to identify patterns and causes of traffic accidents. This process utilizes various statistical methods to analyze crash frequency, severity, and types, along with the conditions under which they occur, such as weather, lighting, and road geometries. By understanding these factors, engineers and planners can design targeted interventions to mitigate risks. For instance, the identification of a high incidence of pedestrian collisions at specific intersections may lead to the implementation of improved crosswalk visibility, pedestrian signals, and traffic calming measures such as speed humps or roundabouts to slow vehicle speeds and enhance safety.

Multimodal planning, on the other hand, focuses on creating transportation systems that accommodate various modes of travel, including walking, cycling, public transit, and personal vehicles. This approach recognizes the importance of providing safe, accessible, and efficient transportation options for all users. By designing infrastructure that supports multiple modes of transportation, cities can reduce reliance on personal vehicles, thereby decreasing traffic congestion, lowering greenhouse gas emissions, and improving air quality. Multimodal planning involves the development of comprehensive networks that connect residential, commercial, and recreational areas with accessible and safe routes for pedestrians and cyclists, as well as the integration of public transit systems that are efficient and reliable.

Sustainable transportation strategies further emphasize the need to design and operate transportation systems in a manner that supports long-term environmental, social, and economic goals. These strategies may include the promotion of electric and hybrid vehicles to reduce emissions, the implementation of green infrastructure practices such as permeable pavements and urban greenways to manage stormwater and enhance urban biodiversity, and the adoption of smart technology to optimize traffic flow and reduce energy consumption.

The application of these principles in transportation planning and design requires a multidisciplinary approach that combines engineering expertise with insights from environmental science, urban planning, and public policy. By prioritizing safety and sustainability, transportation engineers can contribute to the development of communities that are not only safer and more livable but also resilient in the face of changing environmental conditions and evolving mobility needs.

Chapter 14: Construction Engineering

Project Administration

Project Documentation: Contracts and Delivery Methods

Construction contracts serve as the legal backbone of any project, outlining the responsibilities, roles, and rights of all parties involved. They are pivotal in defining the scope of work, the budget, timelines, and the standards to which the work must adhere. Understanding the nuances of these contracts is essential for engineers to ensure projects are completed successfully, within budget, and on time.

Construction Contracts typically fall into several categories, including lump sum, cost-plus, time and materials, unit pricing, and guaranteed maximum price contracts. Each type has its specific use case, benefits, and drawbacks. For instance, lump sum contracts are straightforward, with a fixed total price for all work to be performed, making them suitable for projects with well-defined scopes. Conversely, cost-plus contracts reimburse the contractor for direct and indirect costs and, in some cases, include a fee, either fixed or as a percentage of costs, which can be beneficial for projects where the scope is not well-defined.

Procurement Processes in construction involve the selection and acquisition of goods, services, and work necessary to complete a project. This process includes tendering, where bids are invited from potential contractors and suppliers. The procurement strategy must align with the project's goals, budget, and timeline, and may follow traditional methods, such as design-bid-build (DBB), or more integrated approaches, like design-build (DB).

Design-Bid-Build (DBB) is the traditional project delivery method, where the project owner contracts separately with a designer (or architect/engineer) and a contractor. The design phase is completed before a bid is solicited for the construction phase. This sequential approach allows the owner to have complete control over the design, but it can lead to longer project durations due to the distinct phases.

Design-Build (DB), on the other hand, offers a more streamlined approach by combining the design and construction phases under a single contract. This method can lead to faster project completion times and potentially lower costs, as the design-build team works collaboratively to identify efficiencies and innovations. However, the project owner has less direct oversight of the design, relying on the design-build team to meet the project's objectives.

Each delivery method has its advantages and challenges, and the choice between them depends on the project's specific needs, including complexity, timeline, and budget. DBB might be preferred for projects where the owner wishes to have significant input into the design, while DB could be advantageous for projects requiring fast track completion.

In conclusion, the selection of construction contracts, procurement processes, and delivery methods are critical decisions in the administration of construction projects. These elements define the framework within which projects are executed, directly impacting their success. Engineers and project managers must carefully consider these aspects to align with project goals, ensure legal and financial protections, and facilitate smooth project execution.

Project Management Communication and Coordination

Effective **communication** in project management is the cornerstone of successful project delivery. It involves the clear, concise, and consistent exchange of information among all stakeholders, including project teams, clients, contractors, and suppliers. Establishing a robust communication plan at the outset of a project is crucial. This plan should outline the communication channels, frequency, and formats to be used, ensuring that all parties are informed and aligned with the project objectives and progress. Regular project meetings, status reports, and digital communication platforms can facilitate this ongoing dialogue, enabling timely decision-making and issue resolution.

Change orders represent a significant aspect of project management, often resulting from scope alterations, unforeseen challenges, or stakeholder requests. Managing change orders effectively requires a structured process that includes the documentation of the original project scope, detailed records of the change request, assessment of the impact on the project timeline and budget, and formal approval by all relevant parties. Transparent communication about the implications of change orders is essential to maintain trust and manage expectations. Additionally, incorporating flexibility into project planning can help accommodate necessary changes without significantly derailing the project.

Stakeholder coordination is another critical element, involving the identification, engagement, and management of all individuals and groups with a vested interest in the project. Successful stakeholder coordination ensures that the project meets its intended goals while addressing the concerns and requirements of those affected by its outcome. Strategies for effective stakeholder coordination include stakeholder mapping to identify and understand the influence and interest of each stakeholder, regular engagement meetings, and the development of a stakeholder management plan that outlines strategies for communication and involvement throughout the project lifecycle.

Dispute resolution mechanisms are vital for addressing conflicts that may arise during the course of a construction project. Disputes can stem from a variety of sources, including contract disagreements, delays, cost overruns, and quality issues. Implementing a structured approach to dispute resolution can help prevent conflicts from escalating and adversely affecting the project. This approach may include negotiation, mediation, arbitration, or litigation, depending on the nature and severity of the dispute. Early identification and resolution of disputes through open communication and negotiation are preferable, as they tend to preserve relationships and are generally more cost-effective than formal legal proceedings.

Effective project management in construction engineering hinges on the meticulous planning and execution of these components. By prioritizing clear communication, managing change orders with diligence, coordinating stakeholders strategically, and resolving disputes efficiently, project managers can navigate the complexities of construction projects, ensuring successful outcomes that meet the objectives of all parties involved.

Construction Operations and Methods

Safety Standards and Equipment Selection

Ensuring compliance with Occupational Safety and Health Administration (OSHA) standards is paramount in the construction industry, not only to safeguard the health and well-being of workers but also to maintain project integrity and minimize risks of costly delays and legal issues. OSHA's comprehensive regulations cover a wide array of safety concerns, including but not limited to fall protection, scaffolding safety, electrical safety, and personal protective equipment (PPE) requirements. Adherence to these standards necessitates a thorough understanding of the specific regulations applicable to each aspect of construction operations and the proactive implementation of safety measures and training programs.

Fall protection remains a critical area of focus, given the prevalence of fall-related incidents in construction. OSHA mandates the use of fall protection systems in situations where workers are exposed to falls of six feet or more above a lower level. This includes the installation of guardrail systems, safety net systems, or personal fall arrest systems, depending on the nature of the work and the work environment. The selection of an appropriate fall protection system requires an analysis of the work conditions, including the height at which the work is performed, the mobility of the workers, and the feasibility of installing various types of fall protection measures.

Scaffolding safety is another significant concern addressed by OSHA standards, which stipulate requirements for the design, construction, and use of scaffold systems to ensure they are safe and stable. Compliance involves ensuring that scaffolds are capable of supporting at least four times the maximum intended load, are equipped with guardrails and toe boards, and are inspected by a competent person before each use. The choice of scaffolding material—whether wood, metal, or composite—must align with the specific load requirements and environmental conditions of the project.

Electrical safety standards are designed to protect workers from electrical hazards, including electrocution, electric shock, and arc flash. Compliance with these standards involves the implementation of lockout/tagout procedures to ensure that electrical equipment is de-energized before maintenance or repair work is performed. Additionally, workers must be provided with appropriate PPE, such as insulated gloves and protective clothing, and must be trained to recognize and avoid electrical hazards.

The selection of personal protective equipment (PPE) is guided by a hazard assessment of the work environment to identify potential risks to workers' safety and health. Based on this assessment, appropriate PPE must be provided at no cost to workers. This may include hard hats, safety glasses, hearing protection, respirators, and protective clothing. The effectiveness of PPE in mitigating workplace hazards is contingent upon proper selection, fit, and use, underscoring the importance of comprehensive training for workers on the correct use and maintenance of their PPE.

Efficient equipment selection extends beyond PPE to encompass all tools and machinery used on the construction site. This entails evaluating equipment based on its safety features, reliability, and compliance with applicable OSHA standards. For instance, power tools and machinery must be equipped with guards and safety devices to protect workers from hazards such as flying particles and moving parts. Similarly, vehicles and mobile equipment must have backup alarms and be maintained in good working condition to prevent accidents.

Compliance with OSHA safety standards and the efficient selection of equipment are foundational elements of construction operations and methods. These practices not only fulfill legal obligations but also contribute to the creation of a safe work environment, thereby enhancing productivity, reducing the likelihood of accidents and injuries, and safeguarding the overall success of construction projects.

Site Operations: Structures, Erosion, Productivity

In the realm of construction engineering, site operations are pivotal for the successful execution of projects. This section delves into the critical aspects of **temporary structures**, **erosion control**, and **productivity analysis**, providing a comprehensive understanding essential for the FE Civil Exam preparation.

Temporary Structures play a vital role in supporting construction activities. These structures, including scaffolding, formwork, shoring, and barriers, are designed to facilitate the construction process, ensuring safety and efficiency. Scaffolding provides a secure platform for workers, allowing access to various parts of the construction site. Formwork, the mold into which concrete is poured, is crucial for shaping structures according to design specifications. Shoring supports the sides of excavations and trenches, preventing collapses that could endanger workers and delay projects. Understanding the design principles, load calculations, and safety standards for these temporary structures is essential for civil engineers.

Erosion Control is another critical aspect of site operations, aimed at preventing water or wind from eroding the soil on a construction site. Effective erosion control measures are vital for protecting the site and surrounding environment, maintaining soil stability, and complying with regulatory requirements. Techniques such as silt fencing, sediment basins, and erosion control blankets are commonly employed. Additionally, the strategic placement of vegetation can stabilize soil and absorb water runoff. Civil engineers must be adept at designing and implementing these measures, taking into account the site's topography, soil composition, and expected weather conditions.

Productivity Analysis in construction projects involves evaluating the efficiency of labor, equipment, and processes. This analysis is crucial for identifying areas for improvement, optimizing resource allocation, and minimizing waste. Key performance indicators (KPIs) such as labor productivity rates, equipment utilization, and time to completion are analyzed. Techniques like the Critical Path Method (CPM) and Lean Construction principles can be applied to streamline operations, reduce delays, and enhance overall productivity. Understanding these analytical tools and methodologies is essential for civil engineers to effectively manage and execute construction projects.

Incorporating knowledge of temporary structures, erosion control, and productivity analysis into the FE Civil Exam preparation equips candidates with the practical skills and theoretical understanding necessary for addressing the challenges of construction engineering. Mastery of these topics not only aids in passing the exam but also lays a solid foundation for a successful career in civil engineering.

Project Controls

Scheduling and Tracking Techniques

Gantt Charts are a pivotal tool in project scheduling and tracking, offering a visual representation of the project timeline and the duration of each task. They display the start and end dates of the individual tasks and the overall project. This allows project managers to see the overlap of tasks and understand which tasks are dependent on others to start or finish. The horizontal bars in a Gantt chart represent task duration, making it easier to identify critical milestones and adjust timelines as needed.

Critical Path Method (CPM) is a step further into project management sophistication. It involves identifying the longest stretch of dependent activities and measuring the time required to complete them from start to finish. This is known as the critical path. Calculating the critical path enables project managers to identify which tasks directly impact the project completion time. The formula for calculating the critical path involves identifying the earliest start (ES), earliest finish (EF), latest start (LS), and latest finish (LF) times for each task. The critical path can be determined by applying the formula:

$$EF = ES + Duration$$

$$LS = LF - Duration$$

Where $Duration$ is the time required to complete a task. Tasks on the critical path have no slack time, meaning any delay in these tasks will directly affect the project's completion date.

Resource Allocation is the distribution of available resources, including manpower, equipment, and materials, among the various tasks of a project. Effective resource allocation ensures that each task has the necessary resources to be completed on time without overallocation or underutilization, which can lead to project delays or increased costs. Tools like resource leveling and resource smoothing are used in conjunction with Gantt charts and CPM to optimize the allocation of resources over the project duration.

Activity Relationships are essential in understanding how tasks interact with one another. There are four primary types of activity relationships in project management:

1. **Finish-to-Start (FS):** A task must finish before the next one can start, the most common relationship.
2. **Start-to-Start (SS):** A task must start before or simultaneously with another.

3. **Finish-to-Finish (FF):** A task must finish simultaneously with another.
4. **Start-to-Finish (SF):** A task must start before another can finish, the least common relationship.

Understanding these relationships is crucial for developing a realistic project schedule and applying the Critical Path Method effectively. They help in identifying dependencies between tasks, which is essential for determining the project's critical path and for effective resource allocation.

Incorporating Gantt charts, CPM, resource allocation strategies, and a thorough understanding of activity relationships into project controls enables civil engineers to manage construction projects more efficiently. This comprehensive approach ensures that projects are completed on time, within budget, and according to specifications, thereby enhancing project success rates and contributing to career advancement in the field of civil engineering.

Performance Metrics in Earned Value Analysis

Earned Value Analysis (EVA) is a comprehensive method to track project performance and progress by combining measurements of scope, schedule, and cost in a single integrated system. EVA is predicated on the principle that the value of work performed is indicative of the project's actual progress. The key components of EVA include:

- **Planned Value (PV)**, also known as Budgeted Cost of Work Scheduled (BCWS), represents the total cost of work planned to be completed by a certain date.
- **Actual Cost (AC)**, or Actual Cost of Work Performed (ACWP), is the total cost incurred for the work completed by the specified date.
- **Earned Value (EV)**, or Budgeted Cost of Work Performed (BCWP), reflects the value of work actually completed by the specified date.

The fundamental EVA equations are:

$$EV = (Rate\,of\,Work\,Completion) \times (Total\,Project\,Budget)$$

$$PV = (Planned\,\%\,of\,Work\,Completion) \times (Total\,Project\,Budget)$$

$$AC = Total\,Actual\,Cost\,of\,Work\,Performed$$

From these values, two critical performance indicators can be derived:

1. **Cost Variance (CV)**, which is the difference between the earned value and the actual cost:

$$CV = EV - AC$$

A positive CV indicates that the project is under budget, while a negative CV suggests it is over budget.

2. **Schedule Variance (SV)**, which is the difference between the earned value and the planned value:

$$SV = EV - PV$$

A positive SV indicates that the project is ahead of schedule, whereas a negative SV indicates a delay.

Schedule Performance Index (SPI) and **Cost Performance Index (CPI)** are two ratios that provide a quick snapshot of project health:

- **SPI** = $\frac{EV}{PV}$, where an SPI greater than 1 indicates better than planned performance.

- **CPI** = $\frac{EV}{AC}$, where a CPI greater than 1 signifies that the project is under budget.

These metrics allow project managers to forecast the project's cost at completion and its completion date more accurately. They also facilitate the identification of trends that could impact the project's budget and schedule, enabling proactive adjustments.

To apply EVA effectively, project managers should ensure that all project baselines are accurately defined and that changes to scope, schedule, and cost are properly managed and reflected in the project plan. Regular monitoring of EV, AC, and PV, along with analysis of CV, SV, CPI, and SPI, should be integrated into the project management process to identify variances early and implement corrective actions promptly.

EVA serves not only as a performance measurement tool but also as a project management methodology that emphasizes the integration of scope, schedule, and cost management. By understanding and applying EVA, civil engineers can enhance their ability to manage construction projects efficiently, ensuring that they meet their objectives within the allocated budget and timeframe.

Construction Estimating

Cost Estimation Basics

Quantity takeoff is a critical process in construction cost estimation, involving the comprehensive enumeration and measurement of all materials and items necessary for a project. This meticulous process requires a detailed analysis of construction drawings and specifications to identify every material required. The quantity takeoff must be accurate; an underestimation can lead to budget overruns, while an overestimation can inflate the project cost unnecessarily. The quantities of materials such as concrete, steel, wood, and finishes are measured in their respective units—cubic yards for concrete, pounds or tons for steel, board feet for lumber, and square feet for finishes. These measurements are then compiled into a bill of quantities, which serves as a foundation for costing and procurement.

Unit rates, on the other hand, represent the cost per unit of measurement of materials or labor necessary to install or construct a project component. These rates are derived from historical data, market research, and supplier quotes, and they must include all costs associated with the delivery and installation of materials, including waste and overheads. For instance, the unit rate for pouring concrete would include the cost of raw materials (cement, aggregate, water), transportation, labor for mixing and pouring, formwork, and any equipment needed. The formula for calculating the cost of a material or task is given by:

$$\text{Total Cost} = \text{Quantity} \times \text{Unit Rate}$$

Labor cost determination is another pivotal aspect of construction cost estimation. It involves assessing the amount of work a laborer can accomplish in an hour and the wage rate. Labor productivity rates vary significantly based on the task, experience of the workforce, site conditions, and the tools and equipment available. For example, the productivity rate for laying bricks is measured in bricks per hour, which can be influenced by the complexity of the pattern and the conditions under which the work is performed. Labor costs must also account for indirect costs such as taxes, insurance, and benefits. The total labor cost for a task can be calculated using the formula:

$$\text{Total Labor Cost} = (\text{Labor Productivity Rate} \times \text{Total Quantity of Work}) \times \text{Wage Rate}$$

Incorporating these elements—quantity takeoff, unit rates, and labor cost determination—into the cost estimation process allows construction engineers to create detailed, precise, and dependable project budgets. This approach enhances financial planning and control while also aiding the competitive bidding process by ensuring that estimates are based on realistic evaluations of project costs.

Detailed Estimates: Contingencies and Cost Adjustments

Contingencies in construction estimating are critical for addressing unforeseen costs that arise during the project lifecycle. These costs can result from various factors such as design changes, unexpected site conditions, or fluctuations in material prices. To calculate a contingency budget, a percentage of the total project cost is typically allocated based on the project's complexity and risk level. This percentage can vary but often ranges from 5% to 15%. The formula for calculating the contingency budget is:

$$\text{Contingency Budget} = \text{Total Project Cost} \times \text{Contingency Percentage}$$

It's essential for construction engineers to assess the project's specific risks and adjust the contingency percentage accordingly to ensure it adequately covers potential unforeseen expenses without significantly inflating the overall project cost.

Overhead costs in construction projects encompass the expenses related to general project administration, site management, and support services that are not directly tied to a specific project task. These costs include salaries of the project management team, office expenses, utilities, and equipment rental. Overhead costs are typically calculated as a percentage of the total direct costs (labor, materials, and subcontractor costs). The formula for determining overhead costs is:

$$\text{Overhead Costs} = \text{Total Direct Costs} \times \text{Overhead Percentage}$$

The overhead percentage can vary significantly depending on the company's operational efficiency and the project's duration and complexity. Accurate calculation and allocation of overhead costs are crucial for ensuring the project's financial feasibility and competitiveness in bidding processes.

Cost adjustments in project bidding are necessary to account for changes in project scope, labor rates, material costs, and market conditions. These adjustments ensure the bid reflects the current and anticipated costs accurately. Factors influencing cost adjustments include inflation rates, availability of materials, labor market conditions, and regulatory changes. To adjust costs effectively, construction engineers should monitor market trends and use current data for estimating material and labor costs. Additionally, leveraging historical project data can provide insights into typical cost variances and help refine future estimates.

Adjusting for future cost fluctuations involves applying an escalation rate to the estimated costs, calculated based on historical inflation rates and market forecasts. The formula for applying an escalation rate is:

$$\text{Adjusted Cost} = \text{Estimated Cost} \times (1 + \text{Escalation Rate})$$

This approach helps in creating a more accurate and competitive bid by anticipating and incorporating potential increases in costs over the project's duration.

Incorporating contingencies, accurately calculating overhead, and making informed cost adjustments are fundamental aspects of detailed construction estimating. These practices enable construction engineers to develop comprehensive, realistic, and competitive bids, ensuring financial viability and project success. By meticulously addressing these factors, engineers safeguard against budget overruns and enhance their ability to deliver projects within the established financial parameters, thereby contributing to the firm's reputation and financial health.

Interpretation of Engineering Drawings

Plan Reading: Understanding Construction Documents

Understanding construction plans, sections, elevations, and details requires a comprehensive grasp of the visual language used in engineering and architectural drawings. These documents serve as a blueprint for the construction process, detailing every aspect of the project from the ground up. The ability to accurately interpret these drawings is crucial for engineers, as it ensures that the envisioned design is translated into physical reality with precision and accuracy.

Construction plans, often referred to as plan views, are drawn from a bird's-eye view, providing a comprehensive layout of the project. These plans detail the locations of walls, windows, doors, and other features relative to the project's overall layout. It's essential to understand the scale at which these plans are drawn, as this influences the interpretation of dimensions and distances between features. The scale is typically noted in the legend or title block of the drawing, ensuring that measurements are accurately conveyed and understood.

Sections, on the other hand, offer a cut-through perspective of the construction, providing a view that cannot be seen from the plan's layout. These are particularly useful for understanding the relationships between different levels of a structure, such as the foundation, floors, and roofing. Sections can reveal the construction details of these elements, including the materials to be used, the thickness of walls, and the type of insulation. By analyzing section drawings, engineers can assess the structural integrity and functional aspects of the building, such as ventilation, drainage, and thermal performance.

Elevations are orthogonal projections that depict the exterior faces of the building. These drawings are crucial for understanding the aesthetic and functional aspects of the project, including the arrangement of windows, doors, and other architectural elements. Elevations also provide information on the materials to be used on the facade, offering insights into the building's appearance upon completion. By examining elevation drawings, engineers can ensure that the project meets both the aesthetic preferences of the client and the functional requirements of the structure.

Details are zoomed-in drawings of specific aspects of the construction, providing a close-up view of complex or unique features. These drawings are essential for understanding how different components of the project fit together and are often accompanied by notes and specifications that describe the materials, finishes, and construction techniques to be used. Detail drawings ensure that every aspect of the construction is executed with precision, from the installation of structural connections to the application of waterproofing membranes.

In interpreting these drawings, it's crucial to be familiar with the standard symbols, notation, and abbreviations used in construction documentation. These conventions allow for the efficient communication of complex information and ensure that the drawings are understood universally by all professionals involved in the construction process. Additionally, understanding the sequence in which these drawings are to be read and applied during construction is vital for the successful execution of the project.

By mastering the interpretation of construction plans, sections, elevations, and details, engineers can ensure that the projects they oversee are built according to the specified standards and requirements. This skill not only facilitates the accurate translation of design concepts into physical structures but also plays a critical role in identifying potential issues before construction begins, saving time, resources, and ensuring the project's success.

Specifications and Symbols in Drawings

Interpreting symbols, notes, and material specifications in engineering drawings is a critical skill for construction engineers, requiring a detailed understanding of the standardized visual language used across the industry. Symbols in construction drawings are essentially a shorthand way to convey complex information about materials, finishes, and construction methods without the need for lengthy descriptions. Each symbol is standardized and universally recognized within the engineering community, ensuring clear and unambiguous communication between all parties involved in a construction project.

Material specifications in drawings detail the requirements for the materials to be used, including their quality, strength, and durability standards. These specifications are crucial for ensuring that the constructed facility meets the intended design life and usage requirements. They often reference specific standards or codes, such as ASTM or ACI, which provide the detailed criteria that materials must meet. Understanding these specifications is essential for engineers to ensure compliance with design requirements and regulatory standards.

Notes on drawings serve as additional instructions or clarifications for the construction team. They may provide details on installation methods, special conditions, or preferences of the design team that are not easily conveyed through symbols or standard drafting practices. Notes can also specify testing or inspection requirements to verify that the construction meets the design intent.

The ability to accurately interpret these elements in construction drawings directly impacts the quality, safety, and efficiency of the construction process. Misinterpretation can lead to construction errors, increased costs, and delays. Therefore, engineers must be proficient in reading and understanding every detail conveyed through symbols, notes, and material specifications.

Symbols used in construction drawings can represent a wide range of information, from the type of material (e.g., concrete, steel, wood) to specific components of the building systems (e.g., electrical outlets, plumbing fixtures, HVAC units). For instance, a dashed line might represent an electrical conduit hidden within a wall, while a solid line might indicate a structural beam. Similarly, symbols for doors and windows in plan view provide information about their operation, material, and direction of swing, which are critical for both the construction and eventual use of the building.

Material specifications might include references to specific grades of steel or types of concrete mixtures. For example, a note might specify "ASTM A36 steel" for structural members, indicating that the steel used must meet the standards set forth by ASTM for that grade, which includes specific requirements for chemical composition and mechanical properties. Similarly, concrete might be specified with a note such as "4000 psi concrete with 6% air entrainment," which informs the contractor of the required strength and durability characteristics of the concrete to be used.

Notes on the drawings might include instructions such as "Verify dimensions on site before fabrication," which directs the contractor to take site measurements to ensure that prefabricated components will fit as intended. This can be crucial in renovation projects or in situations where as-built conditions may vary from the drawings.

For engineers preparing for the FE Civil Exam, developing a strong proficiency in interpreting these elements of construction drawings is essential. This skill not only aids in passing the exam but also forms the foundation for effective practice in the field of construction engineering. Mastery of this visual language enables engineers to translate design concepts into reality accurately, ensuring that projects are built to specification, within budget, and on schedule.

Conclusion

Acknowledgment and Gratitude

Your decision to invest in this guide reflects a commendable commitment to not only advancing your career but also to the broader goal of contributing to the engineering field with enhanced knowledge and skills. The path to passing the FE Civil Exam is rigorous and demands a deep understanding of a wide range of topics, from the fundamentals of mathematics and statistics to the complexities of structural engineering and environmental considerations. This guide has been meticulously designed to serve as a comprehensive resource, aiming to equip you with the necessary tools and insights to navigate the exam successfully.

The importance of continuous learning in the engineering profession cannot be overstated. The field is ever-evolving, with new technologies, materials, and methodologies constantly emerging. By embracing a mindset of lifelong learning, you position yourself at the forefront of innovation, ready to tackle the challenges of tomorrow with confidence. This guide is but one step in your ongoing journey of professional development. As you move forward, remember that each topic covered, from the intricacies of fluid mechanics to the principles of transportation engineering, builds upon a foundation that supports your growth as an engineer.

In preparing for the FE Civil Exam, you've taken a significant step toward securing not just a credential, but a testament to your dedication, expertise, and readiness to contribute to the engineering community. This achievement will open doors to new opportunities, from career advancement to a broader scope of projects and responsibilities. The knowledge you've gained through your studies extends beyond the exam; it is a valuable asset that you will carry into your professional endeavors, informing your decisions and enhancing your capacity to innovate and solve problems.

As you continue on your path, remember the value of the connections you forge, the experiences you gain, and the lessons you learn. The field of civil engineering is vast and varied, offering endless opportunities for specialization and exploration. Your journey is unique, shaped by your interests, goals, and the challenges you choose to tackle.

Encouragement for Future Growth

The pursuit of further certifications and professional development is a testament to an engineer's commitment to excellence and innovation. The FE Civil Exam is a significant milestone, yet it is just one step in a lifelong journey of learning and professional growth. The engineering landscape is dynamic, with new challenges and technologies emerging at a rapid pace. To remain at the forefront of the industry, continuous education and adaptation are essential.

Professional certifications beyond the FE and PE, such as those offered by the American Society of Civil Engineers (ASCE) and other professional bodies, provide opportunities to specialize and excel in specific areas of civil engineering. These certifications can enhance your expertise in fields like environmental engineering, water resources, structural engineering, and transportation engineering. They signify a higher level of knowledge and experience, often leading to increased job opportunities, leadership roles, and higher salary potential.

Advanced degrees such as a Master's or Doctorate in engineering or related fields can also open new doors in research, academia, and specialized practice. These degrees offer a deeper understanding of complex engineering principles, emerging technologies, and management strategies, equipping you to tackle the most pressing challenges of our time.

Continuing education through workshops, seminars, and online courses is vital for staying updated with the latest industry standards, software tools, and best practices. Many professional organizations and universities offer courses that cover innovative materials, sustainable design practices, and advanced computational techniques. Engaging in these learning opportunities ensures that your skills remain relevant and competitive.

Participation in professional organizations not only provides access to educational resources but also to a network of peers and mentors. These connections can be invaluable for career development, offering insights into industry trends, job openings, and collaborative projects. Moreover, contributing to the engineering community through research, publications, and presentations at conferences can establish you as a thought leader in your field.

In shaping a sustainable and innovative future, engineers must be proactive in their professional development. The knowledge and skills acquired through continuous learning enable engineers to design solutions that improve the quality of life, protect the environment, and drive technological advancement. As you progress in your career, remember that each new certification, degree, or workshop is a building block towards a more sustainable world and a testament to your dedication to the engineering profession.

Made in United States
Cleveland, OH
17 February 2025